ORGANIC CHEMISTRY 2 LECTURE GUIDE 2019

BY RHETT C. SMITH, PH.D.

Marketed by Proton Guru

Find additional online resources and guides at protonguru.com.

There is a lot of online video content to accompany this book at the Proton Guru YouTube Channel! Just go to YouTube and search "Proton Guru Channel" to easily find our content.

Correlating these reactions with your course: The homepage at protonguru.com provides citations to popular text books for further reading on each reaction in this book, so that you can follow along using this book in any course using one of these texts.

Instructors: Free PowerPoint lecture slides to accompany this text can be obtained by emailing IQ@protonguru.com from your accredited institution email account. The homepage at protonguru.com provides a link to citations to popular text books for further reading on each Lesson topic in this primer.

© 2006-2019
Executive Editor: Rhett C. Smith, Ph.D. You can reach him through our office at:
IQ@protonguru.com

Cover photo courtesy of William C. Dennis, Jr.

Printed in the United States of America

10 9 8 7 6 5 4 3 2 1

ISBN 978-0578415017 (IQ-Proton Guru)

Organic Chemistry 2 Lecture Guide 2019

By Rhett C. Smith, Ph.D.

These slides are meant to accompany a course following the sequence found in the "Organic Chemistry 2 Primer 2019" by Houjeiry, Tennyson and Smith. Instructors who have adopted the Primer may obtain these slides in PowerPoint format for use in their course by inquiring at IQ@protonguru.com from an official university email account.

Contents

BONUS Lessons from PART II. Substitution, Elimination and Oxidation

These lessons are also found in Volume 1, "Organic Chemistry 1 Primer" by Smith, Tennyson and Houjeiry. They are repeated here because some courses cover this material in the second semester course.

So far we have only explored alkyl halides as substrates for S_N1, S_N2, E1 or E2 reactions because alkyl halides with Cl, Br or I substituents have the good leaving groups required. If we propose using an alcohol in these reactions:

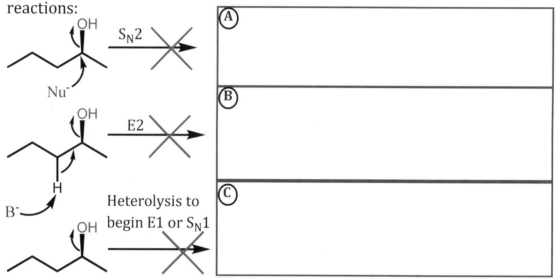

The hydroxide anion is a strong base (unstable anion) and is a bad leaving group. It must be **activated** prior to reaction by these routes.

Notes

One way to convert OH into a good leaving group is to protonate it with a strong acid. This leads to a water leaving group. So, **secondary and tertiary alcohols undergo S_N1 reaction**:

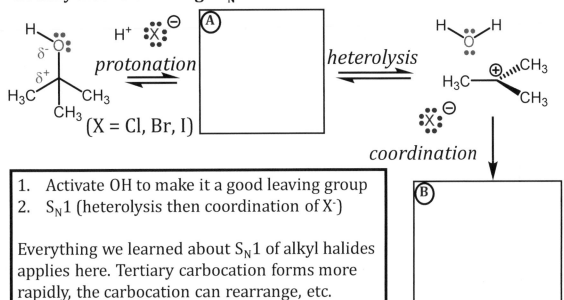

1. Activate OH to make it a good leaving group
2. S_N1 (heterolysis then coordination of X⁻)

Everything we learned about S_N1 of alkyl halides applies here. Tertiary carbocation forms more rapidly, the carbocation can rearrange, etc.

Notes

An activated **methyl, or 1° alcohol will undergo S_N2 reaction** with the halide, which is a good nucleophile:

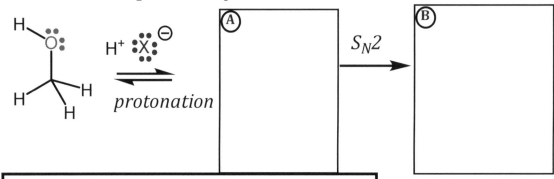

S_N2

1. Activate OH to make it a good leaving group
2. S_N2 (concerted)

Everything we learned about S_N2 of alkyl halides applies here. Walden inversion occurs, less sterically-encumbered substrates react faster, etc.

Notes

Lecture Topic II.11: Substitution and Elimination Reactions of Alcohols
E1 of 2° and 3° ROH: Dehydration

If a 2° or 3° alcohol is activated with an acid having a non-nucleophilic counteranion (i.e., H_2SO_4 or H_3PO_4), an E1 Reaction will occur. Net reaction is loss of water from the alcohol, this particular type of E1 reaction is also called **dehydration**.

1. Activate OH to make it a good leaving group
2. E1 (heterolysis then electrophilic elimination)

Everything we learned about E1 of alkyl halides applies here. Tertiary carbocation forms more rapidly, the carbocation can rearrange, etc.
Note: cannot do E2 (needs strong **base**) of an alcohol with strong acid!

Notes

Another way to activate an OH group is by reaction with PX_3 (X = Cl or Br):

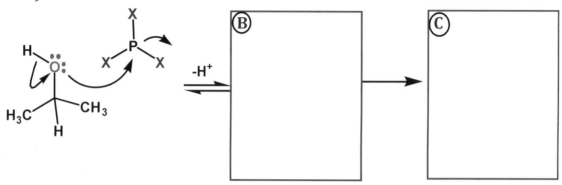

1. Activate leaving group
2. S_N2 reaction – all we have learned about S_N2 applies here; it will not work on a tertiary alcohol, for example.

Note: 1 mole PX_3 can produce 3 mol RX + H_3PO_3

Notes

A third way to activate an OH group is by reaction with thionyl chloride:

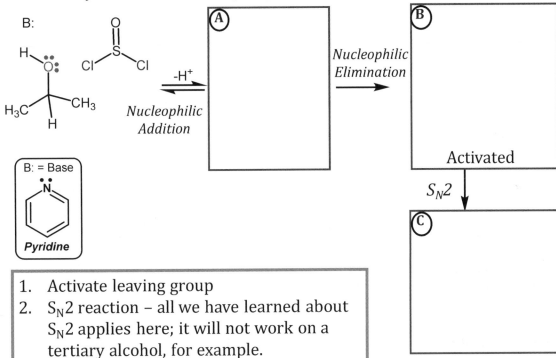

B:

B: = Base

Pyridine

Nucleophilic Addition

-H$^+$

A

Nucleophilic Elimination

B

Activated

S$_N$2

C

1. Activate leaving group
2. S$_N$2 reaction – all we have learned about S$_N$2 applies here; it will not work on a tertiary alcohol, for example.

Notes

Alcohols can also be converted into sulfonate esters. This is not a substitution or elimination reaction, but it is a good way to change the OH into a good leaving group:

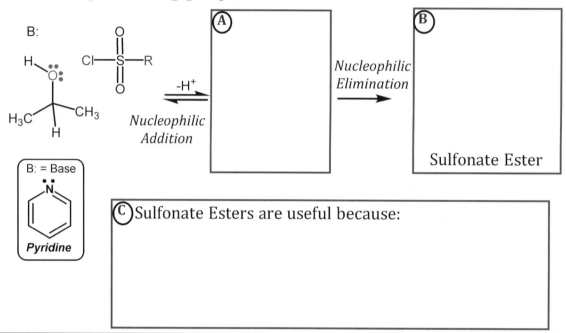

B:

Cl—S—R

-H⁺

Nucleophilic Addition

Ⓐ

Nucleophilic Elimination

Ⓑ

Sulfonate Ester

H₃C   CH₃
    H

B: = Base

N

Pyridine

Ⓒ Sulfonate Esters are useful because:

Notes

The sulfonate is a good leaving group because of resonance stabilization:

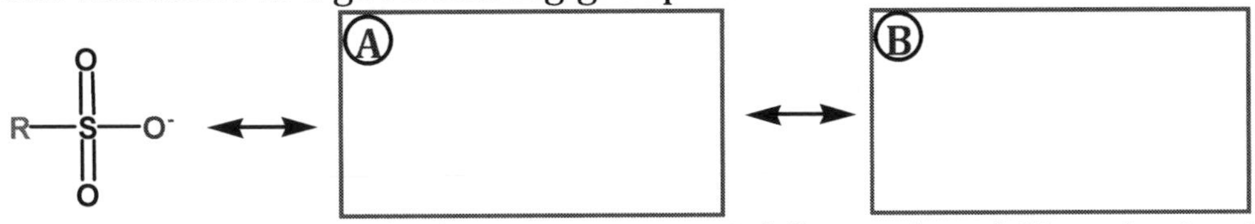

Ⓐ

Ⓑ

A sulfonate ester is ideal starting material for nucleophilic substitution:

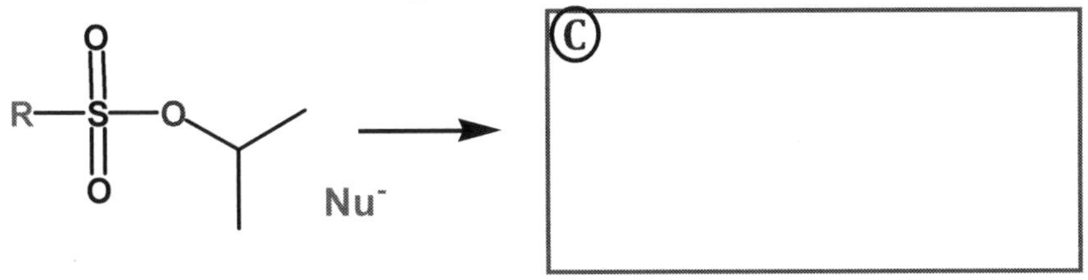

Ⓒ

Nu⁻

Three specific sulfonates are common:

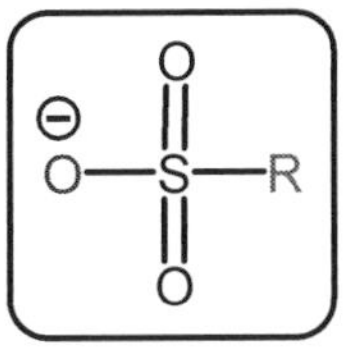

When R =	Anion Name
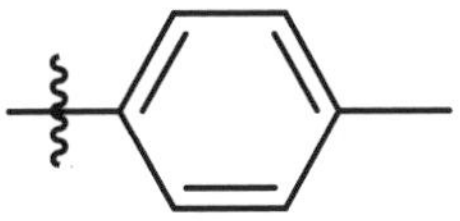 (aromatic ring with CH₃)	Tosylate
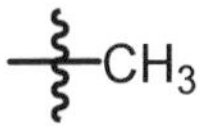 $-CH_3$	Mesylate
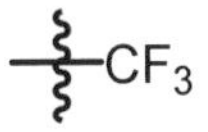 $-CF_3$	Triflate

Notes

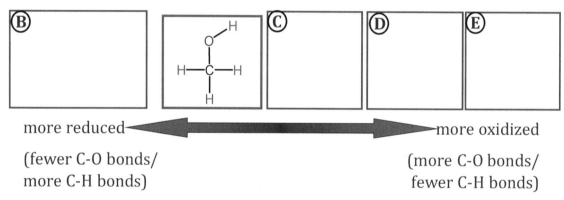

When methane burns in the presence to form carbon dioxide, it is said to undergo "oxidation". The opposite of oxidation is "reduction"

In general, the more C-H bonds that are replaced by C-O (or C to other more electronegative atom) bonds in an alkane, the more oxidized the carbon atom becomes:

more reduced ⟵⟶ more oxidized

(fewer C-O bonds/
more C-H bonds)

(more C-O bonds/
fewer C-H bonds)

Notes

Lecture Topic II.13: Oxidation of Alcohols
Chromium Reagents are Commonly used to Oxidize Alcohols

Oxidation of alcohols is a useful way to make carboxylic acids, aldehydes, and ketones. Strong oxidizing agents like H^+/CrO_4^{2-}, H^+/Cr_2O_7, CrO_3/H_2SO_4 (Jones Oxidation) replace all C-H bonds of an alcohol C with C–O bonds. PCC is weaker and can only replace one C–H with a C–O bond:

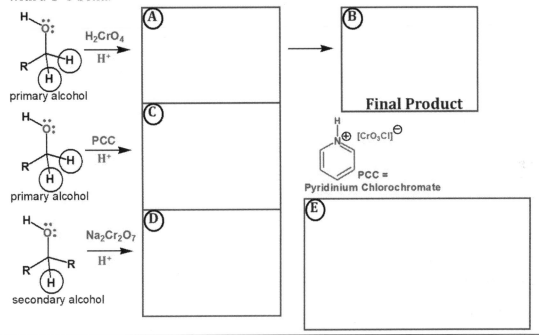

Final Product

PCC = Pyridinium Chlorochromate

Notes

Like alcohols, ethers do not have a good leaving group for substitution reactions. However, the O atom of an ether can be protonated by HX (X = Cl, Br, I) to create a good leaving group. An S$_N$1 or S$_N$2 reaction follows:

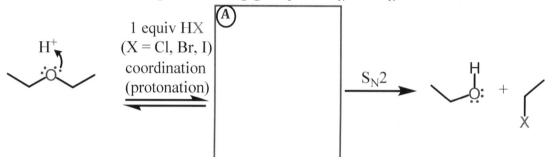

In this example, the ether is symmetric, so it does not matter which side we choose to attack with the nucleophile.

If the carbon to be attacked is methyl or primary:

(B)

Because:

Notes

If the carbon to be attacked by the nucleophile is 2° or 3°:

Ⓐ

This is the case even though a good nucleophile (Cl-, Br- or I-) is present, so this **is a difference from alkyl halides**.

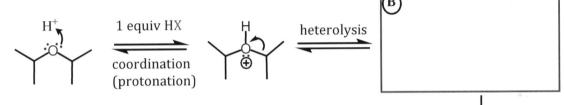

Ⓑ

1. Activate OR to make it a good leaving group
2. S$_N$1 (heterolysis then coordination of X⁻)

Everything we learned about S$_N$1 of alkyl halides applies here. Tertiary carbocation forms more rapidly, the carbocation can rearrange, etc.

coordination

Ⓒ

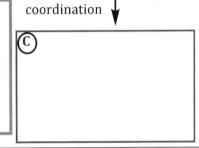

Notes

The ether may not be symmetrical. In these cases:

(A)

If both sides of the ether are capable of S_N1, 3° reacts faster than 2°, as for any other S_N1 reaction. So:

H$^+$ 1 equiv HX

coordination
(protonation)

heterolysis

(B)

coordination

(C)

Notes

An epoxides is a specific type of ether consisting of a 3-membered ring having an oxygen atom in the ring:

Ⓐ

Epoxides are much more reactive than most other ethers because:

Ⓑ

Nucleophiles can thus attack one of the electrophilic carbon atoms, alleviating the ring strain: Ⓒ

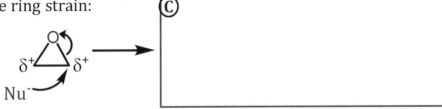

Notes

Under basic (or simply non-acidic) conditions a typical S_N2 reaction occurs, which requires a good nucleophile:

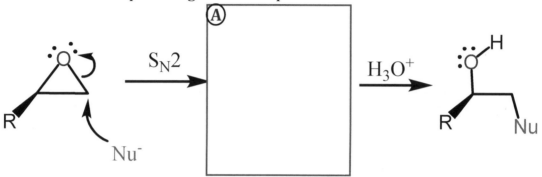

(A)

S_N2

H_3O^+

Nu^-

(B)   Like other S_N2 reactions on neutral substrates, the nucleophile preferentially attacks:

Note that after ring-opening, the oxygen:

Notes

22

Under acidic conditions, the epoxide oxygen is protonated just as in acid cleavage of other ethers:

The less substituted side is attacked unless there is a tertiary site. A tertiary C next to an O with formal charge of +1 has a lot of positive charge, and will be attacked preferentially:

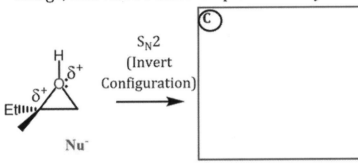

Under acidic conditions, the nucleophile will attack the tertiary C if there is one. Otherwise, the less-hindered side is attacked, just as under basic conditions.

<u>*Notes*</u>

Example. Predict the major product of each reaction, showing stereochemistry where applicable.

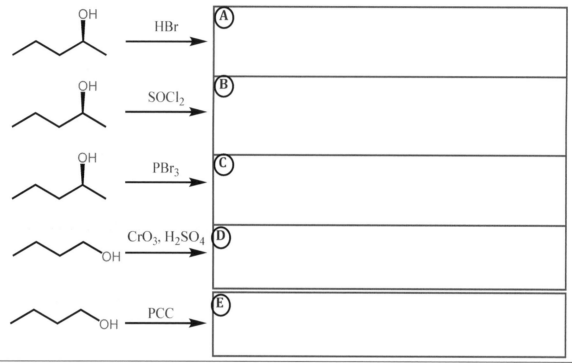

Ⓐ

Ⓑ

Ⓒ

Ⓓ

Ⓔ

Notes

Lecture Topics II.11-15: Recap of Alcohol and Ether Reactivity
Substitution, Elimination and Oxidation

Example. Predict the major product of each reaction, showing stereochemistry where applicable.

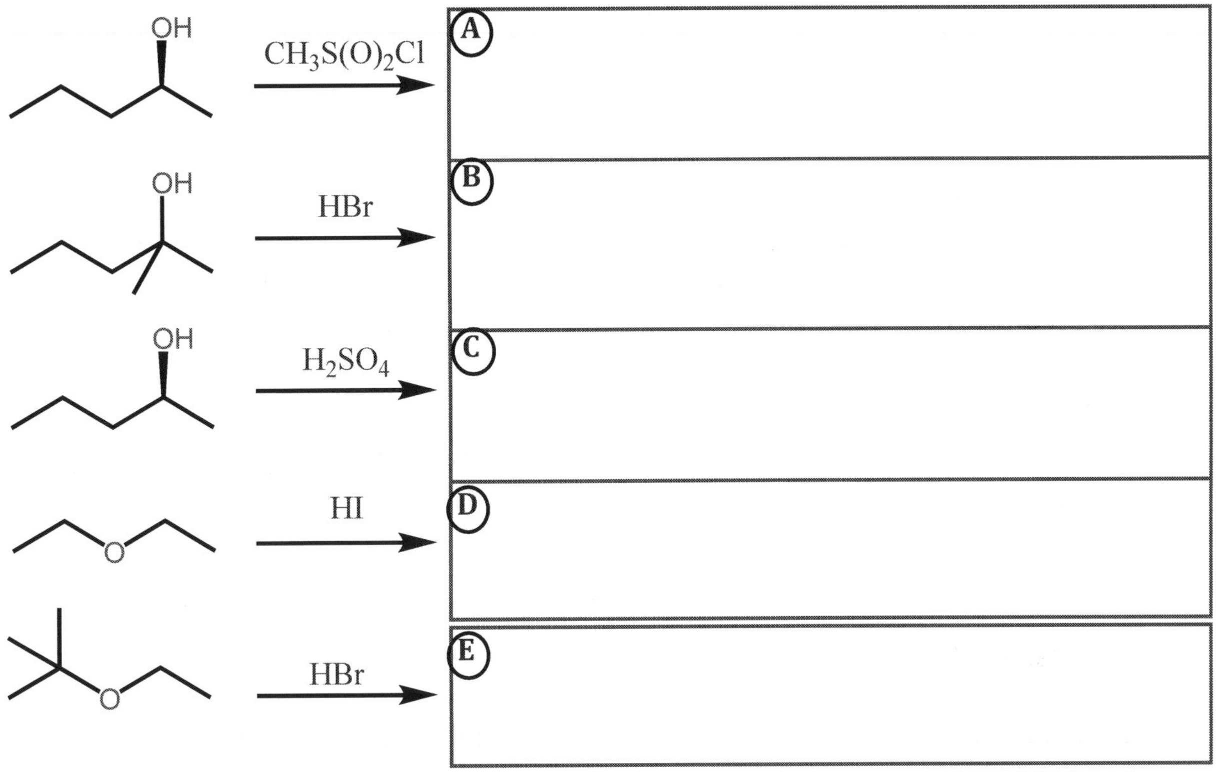

OH

$CH_3S(O)_2Cl$ ⟶ (A)

OH

HBr ⟶ (B)

OH

H_2SO_4 ⟶ (C)

HI ⟶ (D)

HBr ⟶ (E)

Notes

Example. Predict the major product of each reaction, showing stereochemistry where applicable.

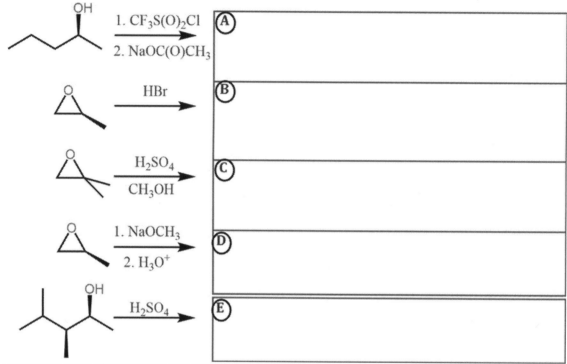

Notes

PART IV: Properties and Reactions of Conjugated and Aromatic Molecules

Lesson IV.1. Pi Conjugation Stabilizes a Molecule
Defining π-conjugation

We know that resonance delocalization stabilizes molecules. We have usually seen this with charged species or radicals, but delocalization can also stabilize neutral species. All that is required are orbitals of appropriate energy and geometry on neighboring atoms. Consider a simple diene in which the two double bonds are separated by one single bond:

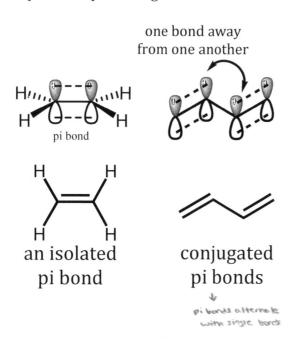

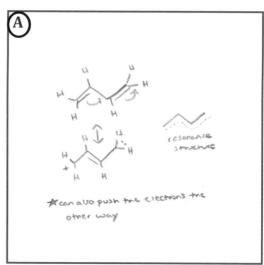

an isolated pi bond

conjugated pi bonds

↓

pi bonds alternate with single bonds

Electrons delocalized via
pi conjugation

Notes

28

When two C=C bonds begin at the same carbon, we get a C=C=C unit:

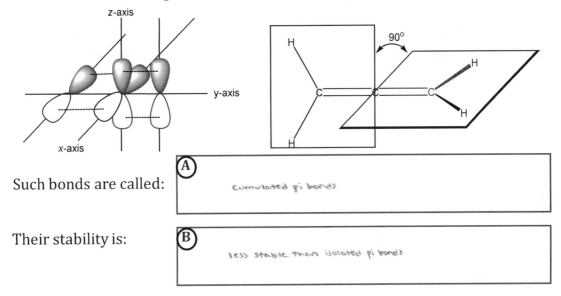

Such bonds are called:

(A) cumulated pi bonds

Their stability is:

(B) less stable than isolated pi bonds

Compared to an isolated C=C

Notes

Lesson IV.1. Pi Conjugation Stabilizes a Molecule
Quantifying Alkene Stability

C=C stability is quantified by measuring:

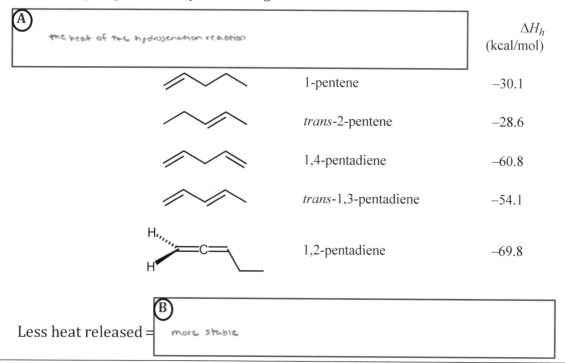

(A) the heat of the hydrogenation reaction

		ΔH_h (kcal/mol)
	1-pentene	−30.1
	trans-2-pentene	−28.6
	1,4-pentadiene	−60.8
	trans-1,3-pentadiene	−54.1
	1,2-pentadiene	−69.8

Less heat released = **(B)** more stable

Notes:

Lesson IV.2. Addition Reactions of Conjugated Dienes
Differing Reactivity

If a C=C bond is π-conjugated, its reactivity may differ from that of a C=C in an isolated alkene:

1-butene $\xrightarrow{\text{1 eq. HX} \atop \text{(X =I, Br, Cl)}}$

markovnikov addition

B

1,3-butadiene $\xrightarrow{\text{1 eq. HX} \atop \text{(X =I, Br, Cl)}}$

? don't know what the product will be

Notes

Lesson IV.2. Addition Reactions of Conjugated Dienes
Identifying Electrophilic Sites

Let us study the process by which the two products form in the case of the conjugated diene:

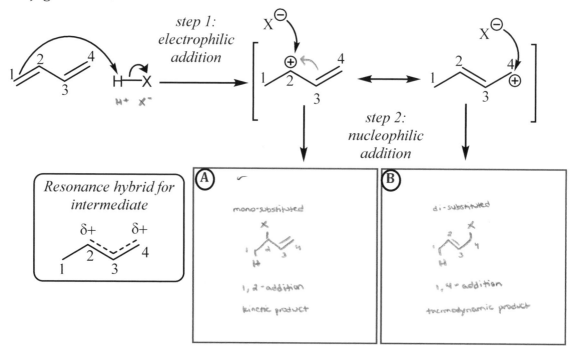

Resonance hybrid for intermediate

(A) mono-substituted

1,2-addition

kinetic product

(B) di-substituted

1,4-addition

thermodynamic product

Notes

• always add the H to carbon number 1

Lesson IV.2. Addition Reactions of Conjugated Dienes
1,2-Addition vs. 1,4-Addition

The 1,2-addition is faster (lower energy of activation, $E_{a1,2}$ in the figure below) because the halide is closer to C2 when it is produced. This is known as a:

(A) proximity effect

The result of this is that the

(B) 1,2-addition

product is formed faster – it is the

(C) kinetic product

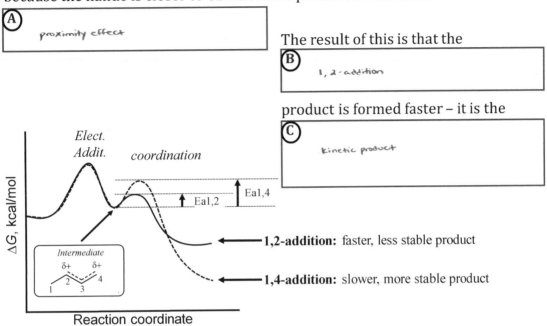

Elect. Addit. *coordination*

ΔG, kcal/mol

$E_{a1,4}$

$E_{a1,2}$

Intermediate

$\delta+$ $\delta+$

1 2 3 4

Reaction coordinate

1,2-addition: faster, less stable product

1,4-addition: slower, more stable product

Notes

The 1,2-addition product is always the kinetic product. However, the (A)

more stable alkene

Is always the **thermodynamic product**. It may result from 1,4-addition (as we saw for 1,3-butadiene), or 1,2-addition, like this:

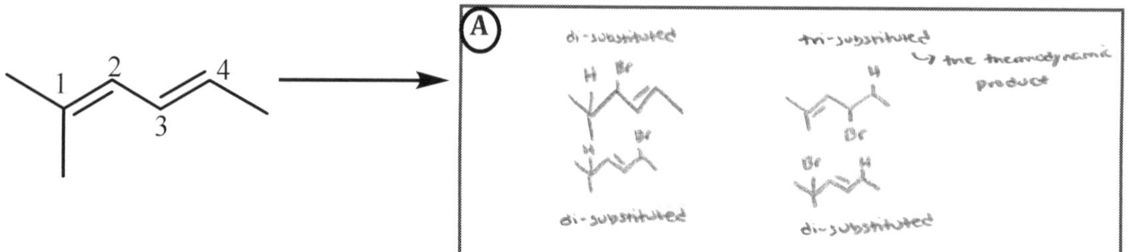

Notes

To reach equilibrium, you need more energy, because the reaction must be able to go from product back to the carbocation intermediate (higher energy barrier than from intermediate to product!).

If the reaction is done at low temperature (*i.e.,* −77 °C) there is not enough thermal energy in the system to reach equilibrium.

0°C or lower

At a higher temperature (*i.e.,* 50 °C) equilibrium can be reached.

Low T:

High T:

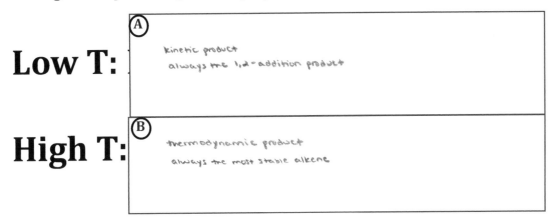

(A) kinetic product
always the 1,2-addition product

(B) thermodynamic product
always the most stable alkene

Notes

Lesson IV.3. Diels-Alder Reaction
A concerted cycloaddition reaction

The Diels-Alder reaction is a [4+2] cycloaddition reaction; it makes a six-membered ring from a diene that supplies 4 atoms and a dienophile that supplies 2 atoms:

This reaction can be classified as a 1,4-addition, because the dienophile adds to the 1 and 4 positions of the diene.

diene   *dienophile*

The mechanism of this reaction is very simple and it is a concerted, pericyclic reaction:

Notes

36

Lesson IV.3. Diels-Alder Reaction
Regiospecificity of Diels-Alder Reactions

One result of the concerted mechanism is that there is no rearrangement of starting materials in the course of the reaction, so the reaction is stereospecific:

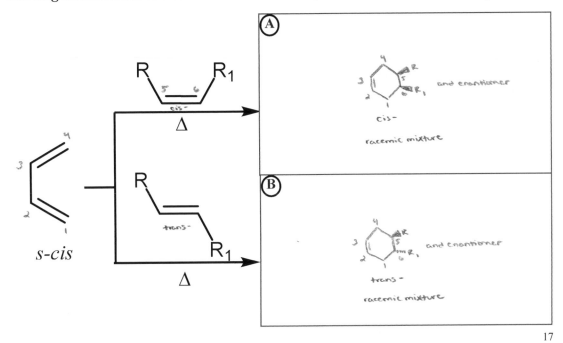

Notes

Lesson IV.3. Diels-Alder Reaction
Electron-Withdrawing Group Effect on Diels-Alder Reaction Rate

Diels –Alder reaction is facilitated by electron-poor dienophiles (dienophile substituted with an electron-withdrawing group):

the reaction is faster
if there is a $\delta+$
near the alkene ->
aka an
electronegative
atom on the carbon

$\delta+$ CN

(A)

CN
and enantiomer

racemic mixture

(B)

Notes

Lesson IV.3. Diels-Alder Reaction

Alkynes and Cyclic Reagents for Diels-Alder Reactions

The dienophile can also be an alkyne:

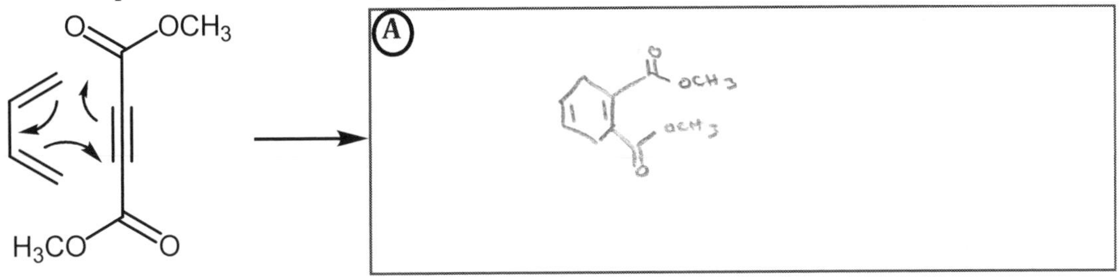

Cyclic compounds can be used as starting materials to create bridged bicyclic systems:

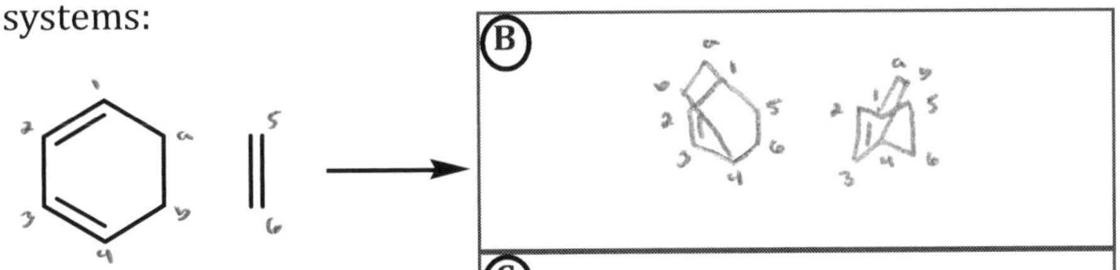

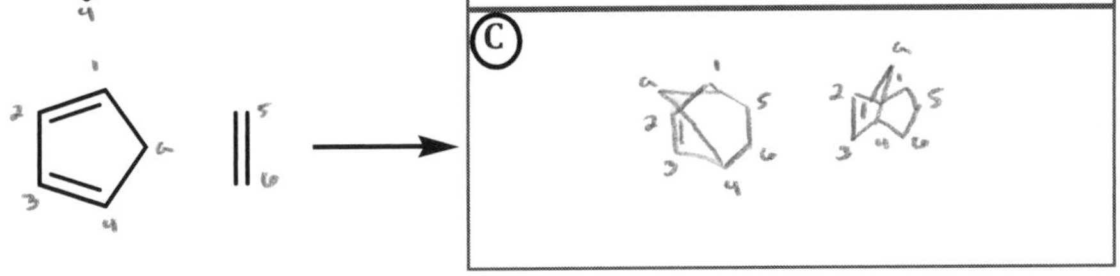

Notes

Lesson IV.4. Aromaticity: A Highly Stabilizing Effect
Benzene is far more stable than an isolated alkene

Description: C_6H_6; six sp^2 hybridized carbons in a planar hexagonal cycle

Observations: <u>C-C-C angles</u>: 120º

                 <u>C-C bonds</u>: 1.40 Å

Structure:

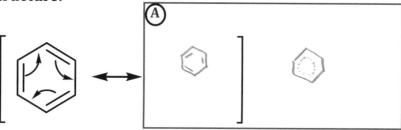

benzene resonance contributors   *resonance hybrid*   *another way to represent benzene*

(B) Pi-electron system:

- gives a great deal of stability to the aromatic system

- therefore, none of the alkene reactions work

Notes

40

In order to exhibit aromaticity, a compound must:

1. cyclic/planar → assume its planar if the other criteria is met

2. resonance around a ring so that every atom in the ring participated → means it must have a double bond, a lone pair, or a positive charge

3. Conform to **Hückel's Rule:**

move 4n+2 electrons in their resonance structure → n has to be an integer

$n = 0$: 2 electrons
$n = 1$: 6 electrons (i.e., benzene)
$n = 2$: 10 electrons

a planar compound with an uninterrupted pi system having **4n electrons is said to be "antiaromatic"**

↳ very unstable

Notes

• nonaromatic compounds are more stable than "antiaromatic" compounds

Examine some compounds to assess whether they are aromatic:

nonaromatic antiaromatic aromatic nonaromatic nonaromatic nonaromatic

<u>Notes</u>

Consider these charged species. Assuming all are planar and have an uninterrupted pi cloud (sp^2-hybridized atoms), check for Hückel's rule condition:

aromatic   antiaromatic   antiaromatic   aromatic   aromatic   nonaromatic

Notes

What if you have a polycyclic system?

 can be aromatic if it follows the criteria

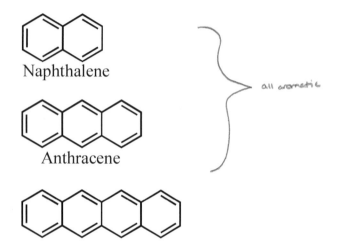

Naphthalene

Anthracene

all aromatic

Notes

What if a ring has heteroatoms (any non-carbon atom) in it?

can be aromatic as well

pyridine

6 electrons to move
aromatic

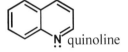

quinoline

10 electrons to move
aromatic

indole

10 electrons to move
aromatic

purine

aromatic

pyrimidine

aromatic

Notes

• if there is a double bond, move it

• only move the lone pair if there is no double bond

Lesson IV.4. Aromaticity: A Highly Stabilizing Effect

Heteroatom-Containing Aromatics

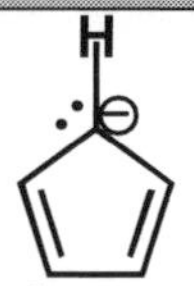

cyclopentadienyl anion

6 electrons to make aromatic

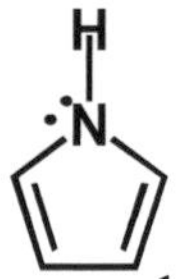

pyrrole

6 electrons to make aromatic

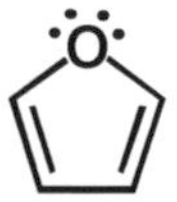

furan

6 electrons to make aromatic

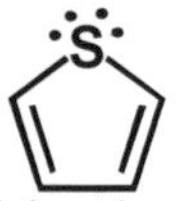

thiophene

6 electrons to make aromatic

***<u>*Molecules tend towards the most stable (lowest energy) state*</u>

<u>*Notes*</u>

The basicity of molecules/sites in a molecule can be predicted by thinking about the behavior of the lone pairs as related to aromaticity:

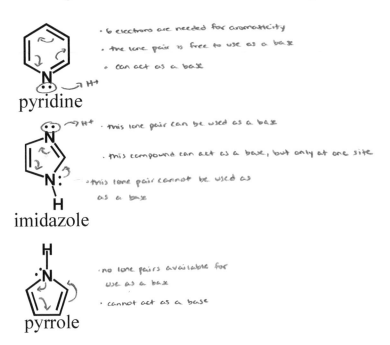

- 6 electrons are needed for aromaticity
- the lone pair is free to use as a base
- can act as a base

pyridine

- this lone pair can be used as a base
- this compound can act as a base, but only at one site
- this lone pair cannot be used as as a base

imidazole

- no lone pairs available for use as a base
- cannot act as a base

pyrrole

Notes

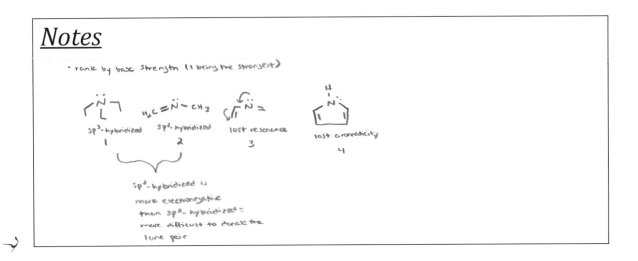

- rank by base strength (1 being the strongest)

sp^3-hybridized sp^2-hybridized   lost resonance   lost aromaticity
1   2   3   4

sp^2-hybridized is more electronegative than sp^3-hybridized = more difficult to donate the lone pair

Likewise, we know that acid strength is tied to the relative stability of the conjugate base anion. If a compound becomes aromatic upon deprotonation, this will make it a stronger acid:

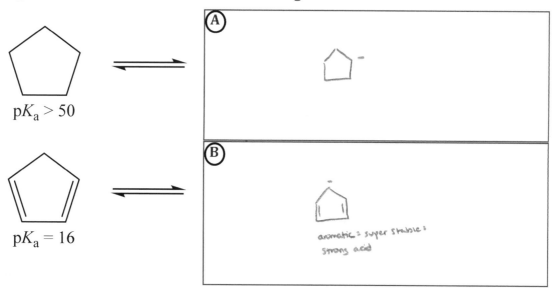

$pK_a > 50$

$pK_a = 16$

(A)

(B)

aromatic = super stable = strong acid

Notes

Many monosubstituted benzene derivatives can be named simply by using benzene as the parent and naming it using rules we learned for alkanes:

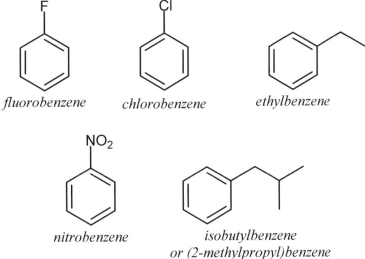

fluorobenzene   *chlorobenzene*   *ethylbenzene*

nitrobenzene   *isobutylbenzene*
or (2-methylpropyl)benzene

When benzene is a substituent, it is called **phenyl** and is abbreviated as —Ph.

Notes

Others have non-systematic names that give some information about the functional group present:

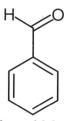 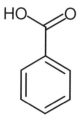 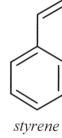 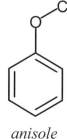 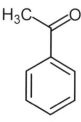

phenol   *benzonitrile*   *benzaldehyde*   *benzoic Acid*   *acetophenone*

Others have well-established common names providing less structural insight:

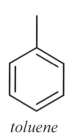

aniline   *styrene*   *anisole*   *toluene*

Notes

50

We have seen that C=C bonds can undergo electrophilic addition with electrophiles to form a carbocation:

(A) make the most stable cation

We have also seen electrophilic elimination of a proton from a carbocation to form a C=C bond:

(B) lose H+   make the most stable alkene

Notes

+ birch reduction: the only reaction we will learn where aromaticity is lost

Li or Na
NH₃(l), ROH
low temperature

· a reduction reaction

For C=C bonds in an arene, a better electrophile is needed than those that do electrophilic addition to isolated C=C bonds. Arenes also do not readily undergo addition reactions like isolated alkenes because this would break the aromaticity:

Instead, arenes undergo **electrophilic aromatic substitution (EAS):**

Ⓐ

all replace hydrogen with an electrophile

Notes

You will learn many electrophilic aromatic substitution reactions;

They all use the same mechanism!

1. Electrophilic addition of the electrophile to the arene (makes a carbocation)
2. Electrophilic elimination of H^+ (yields the aromatic compound with the electrophile on it):

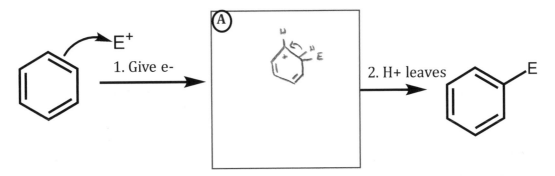

The following pages show specific examples. The only really unique thing is that a different **ELECTROPHILE** is involved in each case, and the reactants you use lead to formation of these electrophiles in different ways...

Notes

The first reaction is called **Friedel-Crafts Alkylation**:

I. The reaction

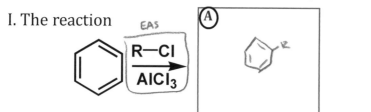

II. How is the electrophile generated?

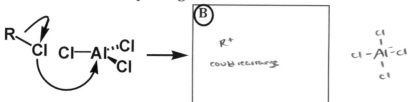

III. The Mechanism:

Notes

When looking at any reaction in which a carbocation intermediate is involved, you must be wary of carbocation rearrangement:

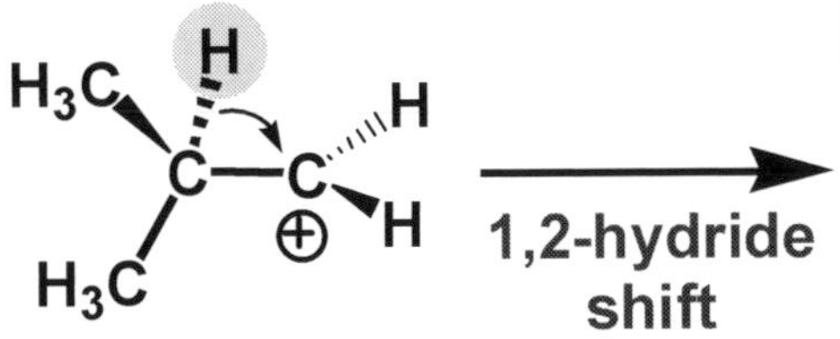

H_3C
H
$C-C$
H
H
$\oplus$
H
H_3C

Primary Carbocation

1,2-hydride shift

(A)

H_3C
R
$C-C$
CH_3
H
$\oplus$
H
H_3C

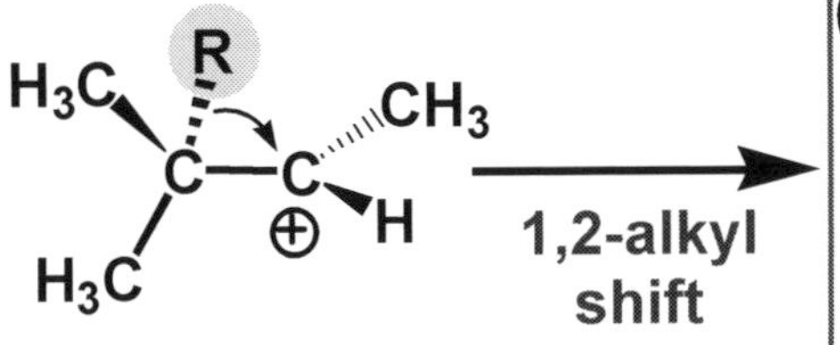

Secondary Carbocation

1,2-alkyl shift

(B)

Notes

Give the major product of the following reaction:

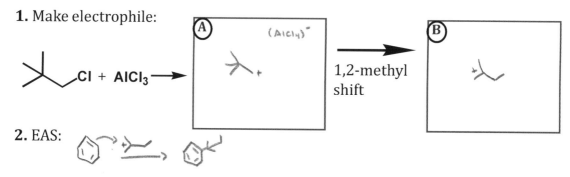

To solve, go through the mechanism:

1. Make electrophile:

2. EAS:

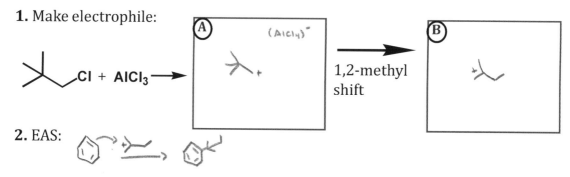

Notes

Lesson IV.7. EAS I: Friedel-Crafts Alkylation and Acylation
Carbocation generation – three ways

The carbocation for use as the electrophile in the Friedel-Crafts alkylation can also be generated by one of the other ways that we already know:

1) Alcohol in the presence of acid (A)

H_2SO_4

2) Alkene in the presence of acid (B)

H_2SO_4

Notes

- may not have to know these

57

So, all three of these reactions are examples of Friedel-Crafts alkylation, just using different ways to make the carbocation needed as the electrophile:

1) Carbocation is from an Alcohol in the presence of Sulfuric Acid

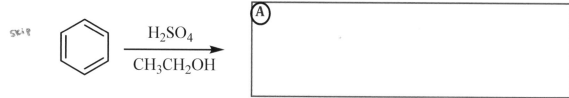

2) Carbocation is from an Alkene in the presence of Acid

3) Carbocation is from Alkyl chloride in the presence of $AlCl_3$

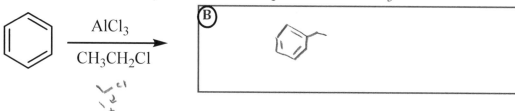

Notes

Friedel-Crafts acylation is another example of EAS. An acyl cation cannot rearrange, so the process is simpler than when a carbocation is involved:

I. The reaction

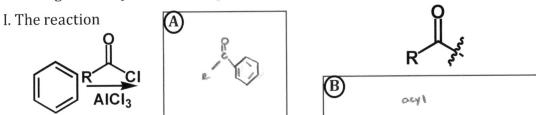

Ⓐ

Ⓑ acyl

II. How is the electrophile generated?

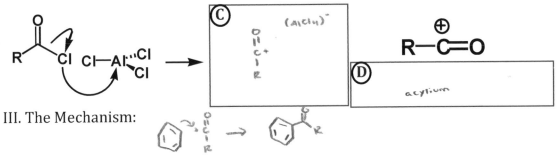

Ⓒ (AlCl₄)⁻

Ⓓ acylium

$R-C\overset{\oplus}{\equiv}O$

III. The Mechanism:

Notes

Sulfuric acid can dehydrate nitric acid to generate a nitronium ion (NO_2^+), which is a very good electrophile:

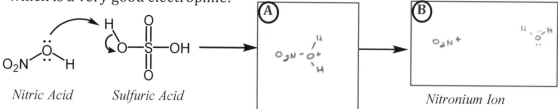

Nitric Acid   *Sulfuric Acid*   *Nitronium Ion*

Once generated, the nitronium ion undergoes the usual EAS process with benzene. This is called a **nitration reaction**:

electrophilic addition → *electrophilic elimination* →

Notes

Lesson IV.8. EAS II: Nitration and Sulfonation of Benzene
Sulfonation

Sulfuric acid can react with itself to generate a sulfonium ion, $[SO_3H]^+$.

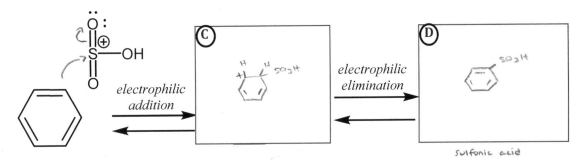

HO_3S Sulfuric Acid   Sulfuric Acid

(A)

(B) sulfonium ion

Once generated, the sulfonium ion undergoes the usual EAS process with benzene. This is called a **sulfonation reaction**:

electrophilic addition

(C)

electrophilic elimination

(D) sulfonic acid

Notes

Sulfur trioxide in sulfuric acid can also be used as the electrophile in a sulfonation reaction. A mixture of SO_3 in sulfuric acid fumes and so it is called **fuming sulfuric acid**. The $-SO_3H$ is a **sulfonic acid** functional group.

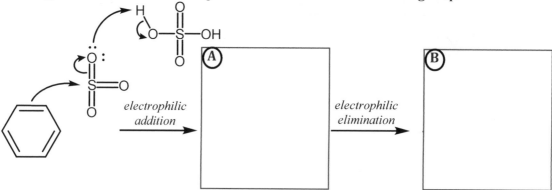

Sulfonation is reversible, so the sulfonic acid group can be removed by heating in water with catalytic acid:

H_2O, trace H_2SO_4, Δ

concentrated H_2SO_4, Δ

<u>*Notes*</u>

Iodenium (I^+) electrophiles can be generated by oxidation of iodine by nitric acid:

$$I_2 + HNO_3 + 2H^+ \longrightarrow 2I^+ + HNO_2 + H_2O$$

Once generated, the iodenium ion undergoes the usual EAS process with benzene. This is called an **iodination reaction**:

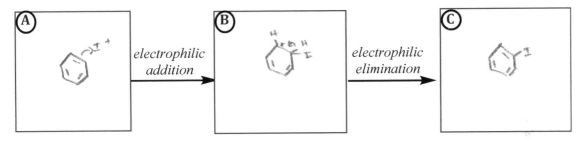

Chlorine and bromine are more electronegative than iodine, so it is more difficult to form bromenium or chlorenium for use in EAS reactions. However, bromine and chlorine can be **polarized** by interaction with iron salts:

$$X\!-\!X \xrightarrow{FeX_3} \overset{\delta+}{X}\!-\!\overset{\delta-}{X}\text{---}FeX_3$$

Notes

The polarized halogens can then be used as electrophiles for EAS with benzene. When X = Cl, this is called a **chlorination reaction**; when X = Br, this is called a **bromination reaction**. Note also that the FeX_3 salt can be generated *in situ* by action of X_2 on Fe metal.

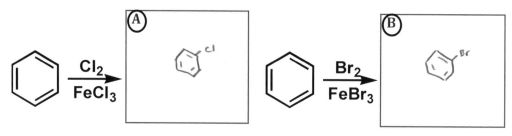

The net reactions are:

Notes

Lesson IV.10. Nomenclature of Polysubstituted Benzene Compounds
Naming polysubstituted benzene and ortho-/meta-/para- system

When naming benzene derivatives, you can use benzene as the parent chain and numbers to denote positions of substituents, like we learned for cycloalkanes in Organic 1. However, there is another widely-used nomenclature method for disubstituted benzenes you must also know:

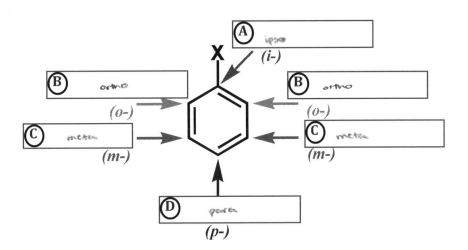

ⓐ ipso
(i-)

ⓑ ortho
(o-)

ⓑ ortho
(o-)

ⓒ meta
(m-)

ⓒ meta
(m-)

ⓓ para
(p-)

Notes

Lesson IV.10. Nomenclature of Polysubstituted Benzene Compounds
Benzene nomenclature examples

(A) 1,2 - dichlorobenzene
or
o - dichlorobenzene

(B) 1- bromo -4- flourobenzene
or
p - bromoflourobenzene

(C) 1- ethyl - 3 -nitrobenzene
m- ethylnitrobenzene

Dimethylbenzene has a common name:

(D) xylene

o-xylene   m-xylene   p-xylene

Notes

Lesson IV.10. Nomenclature of Polysubstituted Benzene Compounds
Benzene nomenclature examples

If a particular molecule contains a benzene derivative with a common name, then use that as the parent, and the substituent that is part of the parent structure always is given the number 1 (it is the *ipso*-position if you are using *o-/m-/p-*):

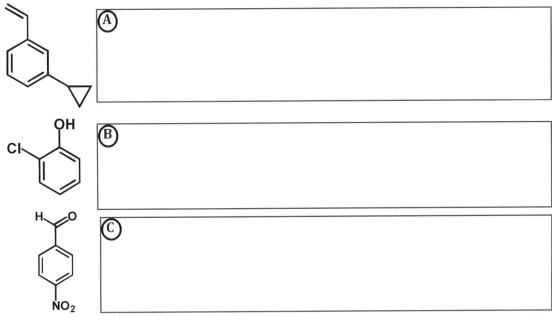

(A)

(B)

(C)

Notes

skip

If there are more than two substituents on the benzene ring, you must number rather than using *o-/m-/p-:*

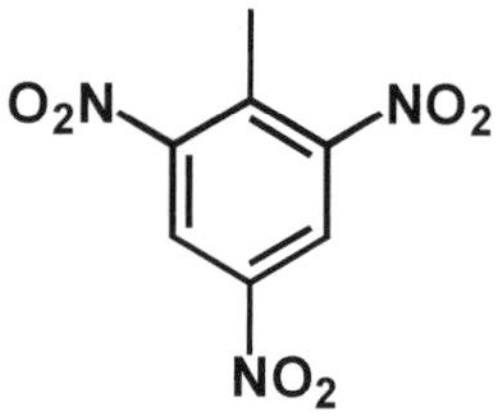

Ⓐ

Ⓑ

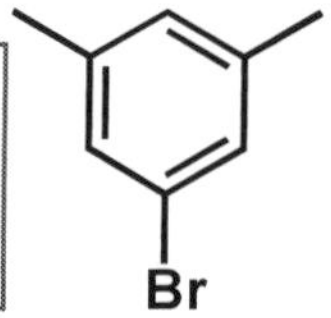

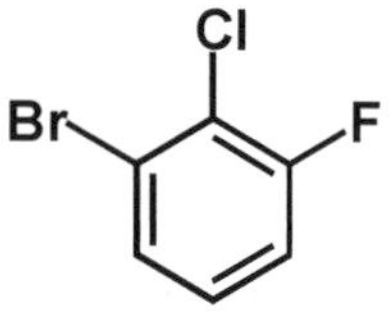

Br

Ⓒ

Notes

Lesson IV.11. Substituent Effects on the Rate of EAS
More stable intermediate = faster reaction

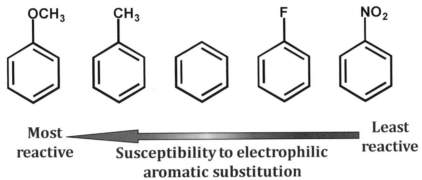

Most reactive ← Susceptibility to electrophilic aromatic substitution → Least reactive

We have a carbocation intermediate, so a more stable carbocation will lead to its more rapid formation.

If we put **electron donor groups** on the benzene ring, it is

A   *faster*

than unsubstituted benzene.

If we put **electron withdrawing groups** on, the arene will be

B   *slower*

than unsubstituted benzene.

<u>Notes</u>

The qualitative reaction coordinate diagram for two EAS reactions reminds us that:

skip

Ⓐ

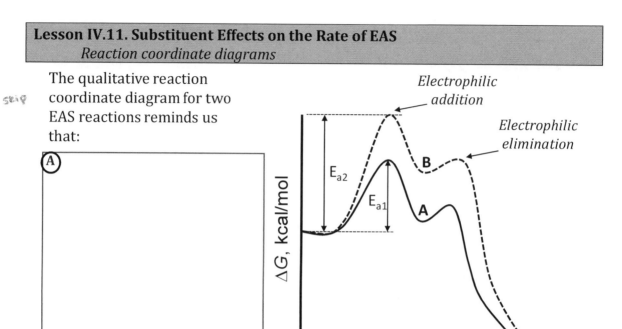

So anything that stabilizes the cation intermediate will speed up the reaction.

Electrophilic addition

Electrophilic elimination

ΔG, kcal/mol

E_{a2}

E_{a1}

B

A

Reaction coordinate

<u>*Notes*</u>

70

Here are some guiding principles for predicting relative activating/deactivating potential of substituents

If the element (X) <u>**DIRECTLY**</u> attached to the benzene ring has a **lone pair**,

 activating group $(\ddot{O}, \ddot{N})$

Except **HALOGENS**:

 slightly deactivating

2. If the substituent is a **HYDROCARBON** (i.e., alkyl, vinyl, or aryl group):

 slightly activating

3. If an element (Y) <u>**ADJACENT**</u> to the directly attached atom is more **electronegative** than carbon or if X has a formal positive charge:

 δ^+ or $+$ by benzene: deactivating group

Notes

So now we have seen the effect of substituents on the names **and** reactivity of aromatic compounds. Recall the trends we observed for substituents on benzene as to the reactivity of the ring in electrophilic aromatic substitution.

We can place substituents into **four general Groups**:

(A)	activating: fastest O-Ö or O-N̈
(B)	slightly activating O-exHy
(C)	slightly deactivating O-X where X= F, Cl, Br, or S
(D)	deactivating (see structures)

Notes

Lesson IV.11. Substituent Effects on the Rate of EAS
Ranking substituent activating ability for EAS

What affects the ability of an element to donate electrons to carbon?

Size – if the element is not in the same row as carbon, it is too big to effectively overlap the pi system.

Electronegativity – because fluorine is so electronegative, it does not donate electrons to the pi system enough to activate the molecule, even though it is the same size.

Combining the rules with these 2 guiding principles, we can group many substituents as activating or deactivating (relative to benzene) for cation formation.

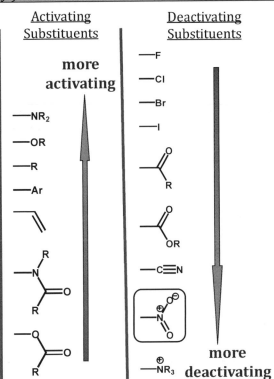

Classify each of the given substituents as being Slightly Activating (SA), Activating (A), Slightly Deactivating (SD), or Deactivating (D), relative to H (plain benzene), with respect to reactivity in Electrophilic aromatic substitution.

(i) CH_3 — SA

(ii) [acetate ester structure] — A

(iii) SO_3H — D

(iv) [N(CH₃)₂ structure] — A

(v) [tert-butyl structure] — SA

(vi) Cl — SD

(vii) [$N^+(CH_3)_3$ structure] — D

(viii) NO_2 — D

(ix) Br — D

(x) [acetyl structure] — D

Notes

Rank these arenes from 1-5 in terms of their reactivity towards electrophilic aromatic substitution, 1 being most reactive and 5 being least reactive. Explain your selections.

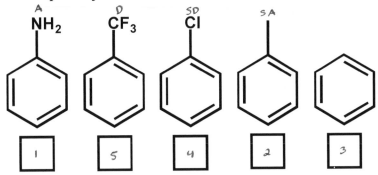

A
NH_2
1

D
CF_3
5

SD
Cl
4

S A
2

3

Notes

If there is already a substituent on a benzene ring and we then attempt to further substitute it, we need to consider what is the major product will be. If a monosubstituted arene is subjected to EAS, we can get *o-*, *m-*, and *p-* products:

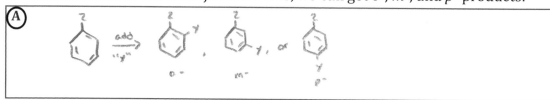

In order to understand which product(s) are formed in highest yield, we must examine the mechanism, keeping in mind one of the **generally applicable principles** used in examining organic chemical transformations:

*EAS = Electrophilic aromatic substitution

Notes

Consider a resonance donor (Z) substituent with electrophile (E) added *o*-, *m*- or *p*-:

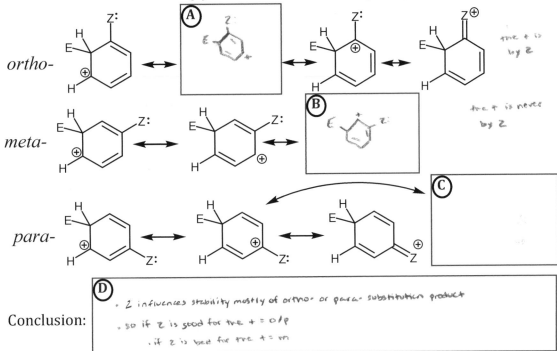

ortho-

the + is
by Z

the + is never
by Z

meta-

para-

Conclusion:

(D)
• Z influences stability mostly of ortho- or para- substitution product
• so if Z is good for the + = o/p
• if Z is bad for the + = m

Notes

Now consider an inductive donor (R = alkyl):

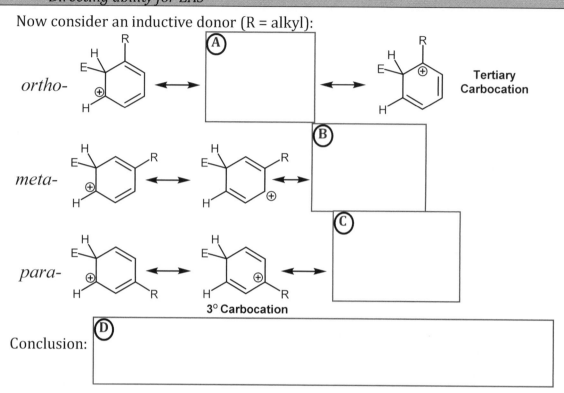

ortho-

Tertiary Carbocation

A

B

meta-

C

para-

3° Carbocation

Conclusion: D

Notes

Next, a halogen (X = F, Cl, Br, I):

ortho-

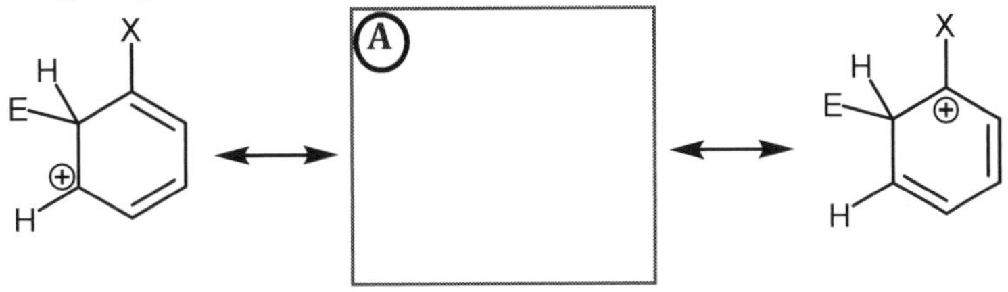

(A)

meta-

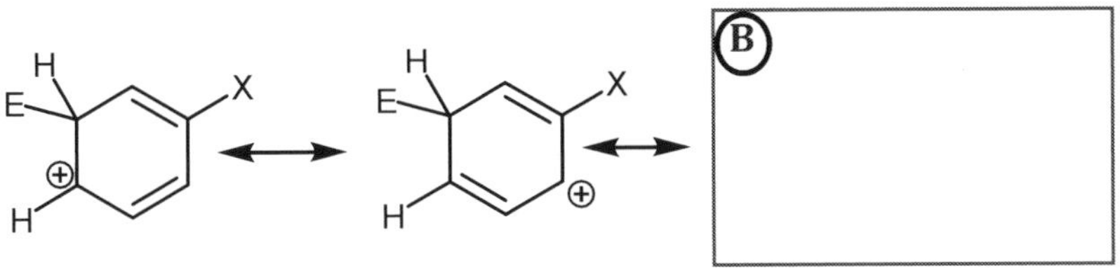

(B)

para-

(C)

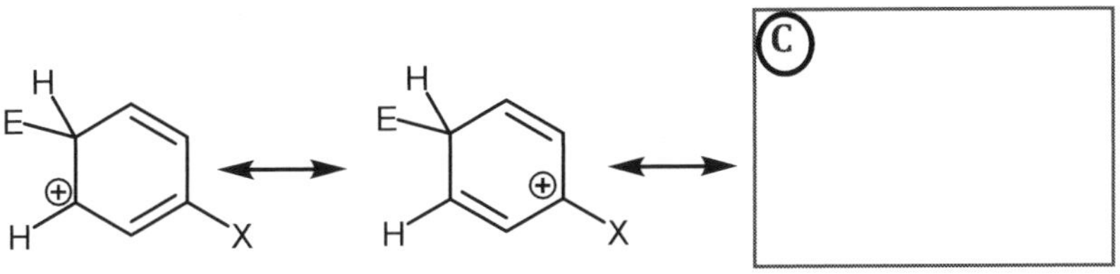

Notes

For halogens consider their interaction with an adjacent carbocation as compared to the hyperconjugation that stabilizes carbocations:

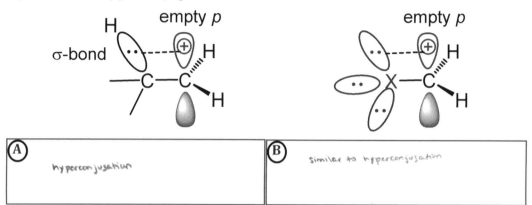

(A)	(B)
hyperconjugation	similar to hyperconjugation

The figure above illustrates that a halogen is able to stabilize a positive charge on an adjacent C. With this knowledge and the resonance contributors on the previous page, we conclude:

(C)

X by e̅ is good for stability

Notes

80

Lesson IV.12. Substituent Effects on the Regiochemistry of EAS
Directing ability for EAS

Finally, consider an inductive withdrawing group (Z):

ortho-

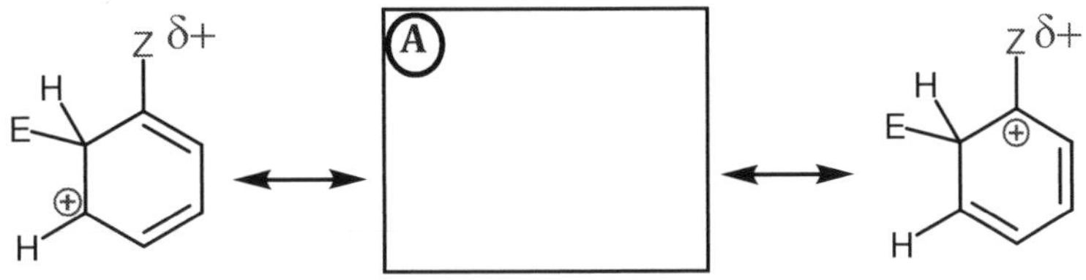

meta-

para-

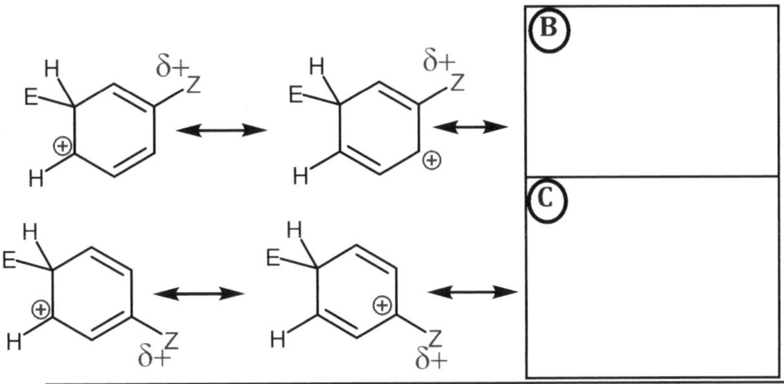

Conclusion: (D)

Notes

A general trend worth noting is that we can correlate a substituent's **EAS Reactivity Group** with its **EAS Directing Type**:

(A)

activating, slightly deactivating, and
slightly deactivating = 2 by + good
so o/p

· deactivating = 2 by + bad so m

and

(B)

Remember that these are *general trends* to help us predict the **major** product; you typically will not get *exclusively* the favored product(s)

Notes

Example: Give the major product(s) of the following reactions using your knowledge of whether the existing substituent is *o-/p-* directing or *m*-directing.

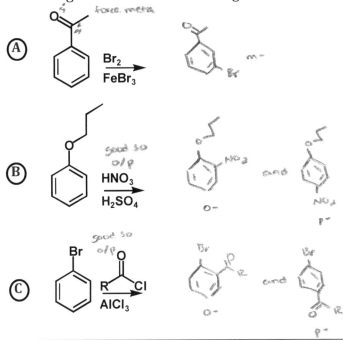

(A) force meta
Br$_2$ / FeBr$_3$ → m-

(B) good to o/p
HNO$_3$ / H$_2$SO$_4$ → o- and p- NO$_2$

(C) good to o/p
R—C(=O)Cl / AlCl$_3$ → o- and p- R

Notes

Lesson IV.13. More on the Directing Effects of Substituents in EAS
Steric considerations for EAS

A complication is that the activating and slightly deactivating substituents direct substitution at both *ortho* and *para* positions. It would be useful to be able to predict how much *o-* and how much *p-* product will be formed. One consideration is:

> **(A)** ortho- is more hindered

Another point to consider is:

> **(B)** 2 ortho- but only one para-

Thus:

> **(C)** generally get both ortho- and para-

<u>Notes</u>

Lesson IV.13. More on the Directing Effects of Substituents in EAS
Steric considerations for EAS Practice

Often, a starting arene will have two substituents on it. In these cases, you need to consider the directing preference of both substituents. As a general rule:

(A) do not place a new substituent between two others if you can help it

You also must consider:

(B)
- Sterics
- if there are two substituents on already, the more activating dominates

Examples:

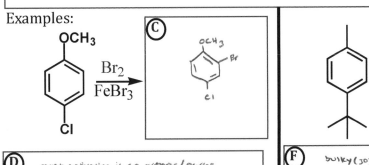

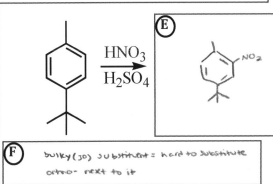

(C)

(E)

Br_2 / $FeBr_3$

HNO_3 / H_2SO_4

(D) most activating is an ortho-/para-director, Br goes ortho-

(F) bulky (3°) substituent = hard to substitute ortho- next to it

Notes

Lesson IV.13. More on the Directing Effects of Substituents in EAS

Multiple Substituents for EAS Practice

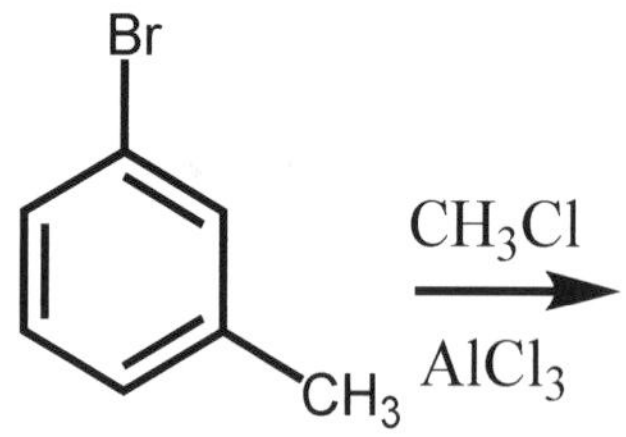

A

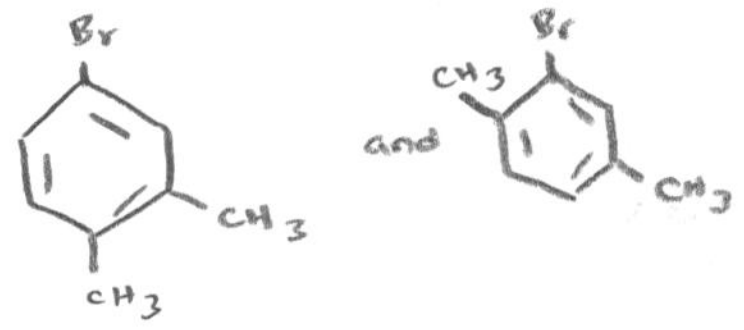

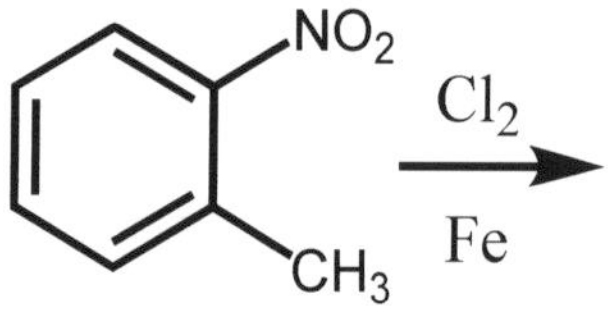

B

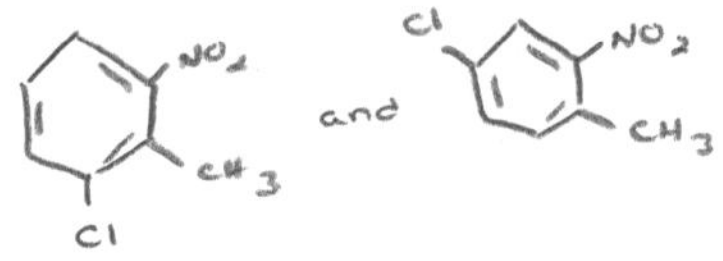

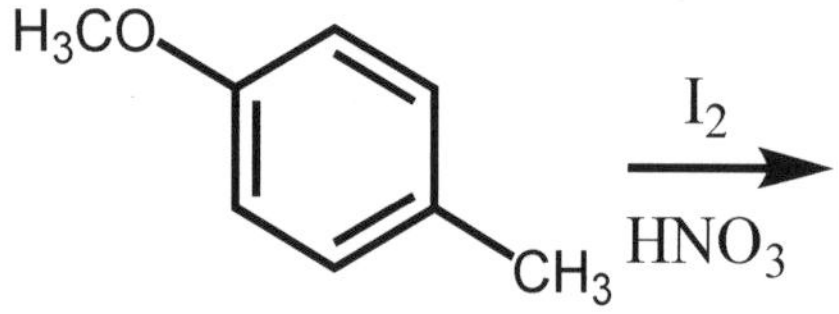

C

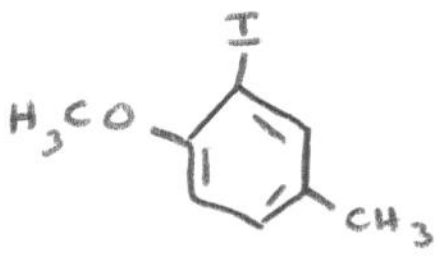

Notes

Lesson IV.14. Oxidation and Reduction of Substituents on Benzene Rings
Alcohol oxidation and pi-bond reduction

In Organic Chemistry,
Oxidation is:

A
- more C-O bonds
  and/or
- fewer C-H bonds

In Organic Chemistry,
Reduction is:

B
- more C-H bonds
  and/or
- fewer C-O bonds

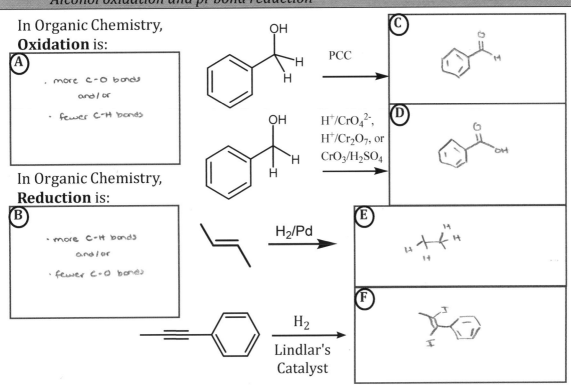

PCC

H^+/CrO_4^{2-},
H^+/Cr_2O_7, or
CrO_3/H_2SO_4

H_2/Pd

H_2
Lindlar's
Catalyst

C

D

E

F

Notes

Note that MnO_2 is an oxidizing agent that will carry out reactions similar to PCC and that $KMnO_4$ will do reactions like the Jones Oxidation.

The benzylic site (site adjacent to benzene ring) has added reactivity:

Any of these starting materials

H^+/CrO_4^{2-}, H^+/Cr_2O_7, CrO_3/H_2SO_4

or 1. dil. $KMnO_4$/base, 2. H_3O^+

(A)

The benzylic carbon must have at least one H for the reaction to proceed:

R is not H

H^+/CrO_4^{2-}, H^+/Cr_2O_7, CrO_3/H_2SO_4

or 1. dil. $KMnO_4$/base, 2. H_3O^+

(B)   no reaction

Notes

Lesson IV.14. Oxidation/Reduction of Substituents on Benzene Rings
Reduction of carbonyls

Several functional groups can undergo reduction if they are adjacent to an aryl ring as well:

(A) *keep the benzene ring intact*

We've seen such a reaction for an alkene:

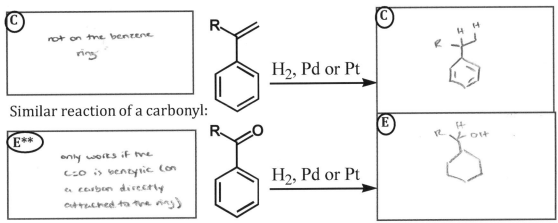

(C) *not on the benzene ring*

R $\quad$ H$_2$, Pd or Pt → (C) *R with H H*

Similar reaction of a carbonyl:

(E**) *only works if the C=O is benzylic (on a carbon directly attached to the ring)*

R $\quad$ H$_2$, Pd or Pt → (E) *R with H OH*

**only works if the C=O is right next to arene

Notes

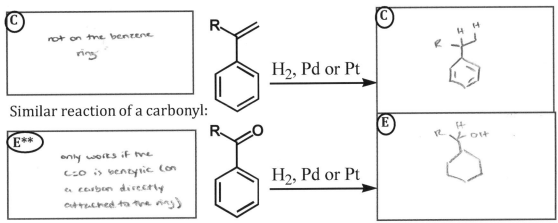

Such conditions also reduce for triple bonds to nitrogen:

A

≡ $\xrightarrow{\text{Pd}, H_2}$

→ benzonitrile

H_2, Pd or Pt →

B

And even to replace O with H in nitro groups:

C

H_2, Pd or Pt →

D

aniline

Another way to do the later reaction:

Sn, HCl →

E

NH_2

aniline

Notes

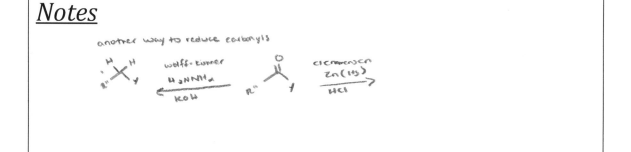

another way to reduce carbonyls

wolff-kishner
H_2NNH_2
KOH

clemmensen
Zn(Hg)
HCl

Lesson IV.15. Radical Halogenation of Allylic and Benzylic Compounds
Radicals abstract hydrogen atoms to make the most stable radical possible

The radical intermediate of radical halogenation is very stable for an allylic or benzylic substrate, but high concentration of halogens can lead to electrophilic addition of X_2 to the double bond in an allylic case. For this reason, special reagents (i.e., NBS) capable of producing only a low concentration of halogen are typically used in such cases:

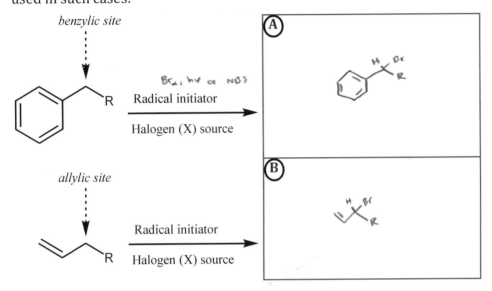

benzylic site

allylic site

Br_2, hv or NBS
Radical initiator
Halogen (X) source

Radical initiator
Halogen (X) source

Ⓐ

Ⓑ

Notes

Lesson IV.15. Radical Halogenation of Allylic and Benzylic Compounds
NBS is a source of dilute bromine

When *N*-bromosuccinimide (NBS) is heated, a low concentration of Br_2 is generated. The initial radicals are generated by heating benzoyl peroxide $(PhC(O)))_2$. These radicals go on to initiate formation of bromine radicals. Once bromine radicals are generated, they can propagate further reaction:

This reaction works if the $-C(H)=CH_2$ is replaced with an arene ring as well.

Notes

Lesson IV.15. Radical Halogenation of Allylic and Benzylic Compounds
Radical bromination practice

Remember that allylic radicals have more than one resonance contributor, such that the actual molecule (hybrid) has radical character at two carbons; the situation becomes even more complicated when there are two allylic sites from which to choose in the starting material:

How does one account for all of these products?

(A)	(B)

Notes

We have seen that electrophiles can do EAS reactions of Arenes. Now we will briefly explore how aromatic compounds react with **Nucleophiles** instead of electrophiles. Arenes may react with nucleophiles via:

(A) S$_N$Ar
  substitution of a leaving group for a nucleophile on
  Sn2   an aromatic

There are some structural requirements for a S$_N$Ar reaction to work:

(B)
• a nitro group with a leaving group that is ortho- or para- to that nitro group

Notes

The net S_NAr reaction is replacement of the substituent that is *ortho-* or *para-* to the strongly withdrawing substituent with the nucleophile:

The S_NAr reaction involves a two-step mechanism:

Ⓑ

· add the nucleophile, then lose the leaving group

Notes

Now that we know the mechanism of the SNAr reaction, we can rationalize the requirement for the leaving group to be *o*- or *p*- to the EWG. Let us consider the nucleophilic addition step:

ortho- nucleophilic addition

anion intermediate is stabilized by EWG

Notes

Consider nucleophilic addition to the site *p*- to the EWG:

para- nucleophilic addition

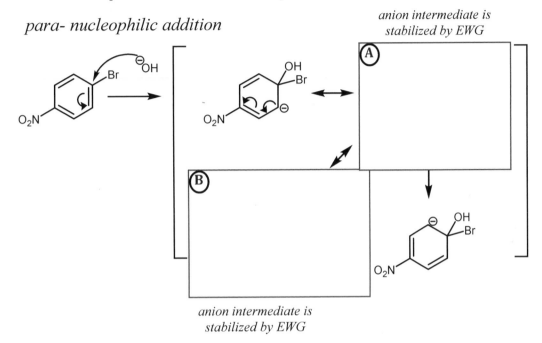

*anion intermediate is
stabilized by EWG*

*anion intermediate is
stabilized by EWG*

Notes

Consider nucleophilic addition to the site *m-* to the EWG:

meta- nucleophilic addition

*anion intermediate
is **NOT stabilized**
by EWG*

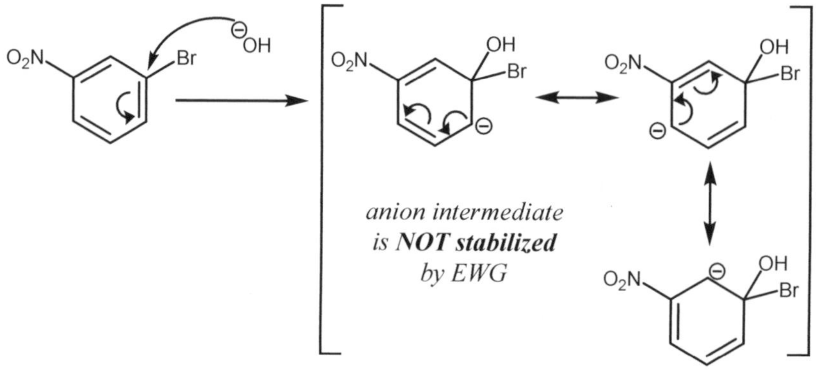

The relative stabilities of the anionic intermediates reveal:

Ⓐ

Notes

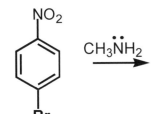

HO$^{\ominus}$

(A)

CH$_3$N̈H$_2$

(B)

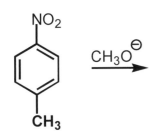

CH$_3$O$^{\ominus}$

(C)

Notes

Diazonium salts $(R-N_2^+)$ are good precursors for transformation into other functional groups on arenes via the following general reaction scheme:

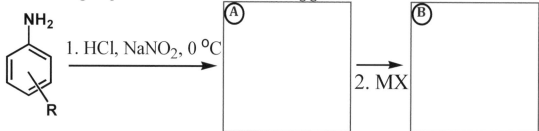

The identity of the group that ends up on the ring depends on what compound MX is added in the second step:

CuCl: X = Ⓒ   CuBr: X = Ⓓ   KI: X = Ⓔ

HBF_4: X = Ⓕ Cu_2O: X = Ⓖ H_3PO_2: X = Ⓗ

Notes

Lesson IV.17. Formation and Reaction of Diazonium Salts
Diazotization application

To illustrate just one case where the diazonium reaction proves very useful, consider an attempted synthesis of this target:

Attempts to synthesize this target using only EAS will quickly prove problematic:

TARGET

Diazotization following EAS will readily give the target:

1. HCl, NaNO$_2$, 0 °C
2. KI

Notes

Give the best synthesis of the following targets from benzene in as many steps as necessary, using the reactions presented in the notes thus far.

I. Some simpler ones:

(A)

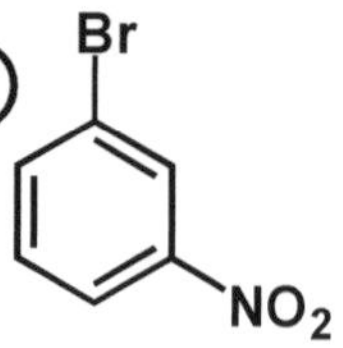

Br

NO$_2$

(B)

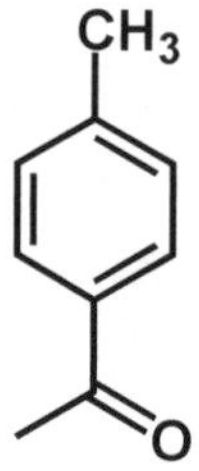

CH$_3$

O

Notes

Summary for Part IV: Putting it all Together
Multistep synthesis

II. Some that require substitution followed by reaction of the functional groups:

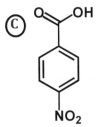

Ⓒ

Ⓓ

NH₂

Br

NO₂

Notes

PART V: Organometallic Compounds and Metal Hydrides

Lesson V.1. Introduction to Organometallics and Metal Hydrides
The need for nucleophilic carbon

In functional groups we studied previously (alcohols, ethers, and alkyl halides) carbon atoms are attacked by nucleophiles. We only know two examples in which a C has a negative charge on it and *acts as a nucleophile*:

A — invert Stereochemistry

1) NaNH$_2$

2) 1-bromobutane

B

Organometallic species feature a C–M bond, so that significant negative charge is on the C atom. This makes organometallic compounds a good source of nucleophilic carbon.

Notes

We should preface our discussion of how organometallic species are prepared by learning a few of the elementary steps that occur between organic species and metals:

Oxidative addition:

$$R\text{—}X + M^n \longrightarrow R\text{—}M^{n+2}$$
$$X = Cl, Br, I \qquad\qquad \overset{|}{X}$$

(A)

$$R\text{—}M$$
$$\overset{|}{X}$$

M is oxidized as the R and X is add

Reductive elimination:

$$R\text{—}M^{n+2} \longrightarrow R\text{—}R' + M^n$$
$$\overset{|}{R'}$$

(B)

reduce M and eliminate some organic from it

Transmetallation:

$$M\text{—}X + M'\text{—}R \rightarrow M\text{—}R + M'\text{—}X$$

(C)

Ligand Exchange:

$$R\text{—}M\text{—}X \rightleftharpoons X\text{—}M\text{—}X$$

(D)

Notes

Lesson V.1. Introduction to Organometallics and Metal Hydrides
Metal hydrides have variable reactivity

Reagents that give H– as nucleophiles (H⁻ is called **Hydride**)

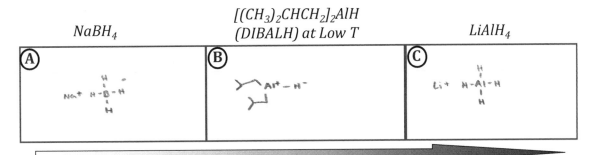

$NaBH_4$

$[(CH_3)_2CHCH_2]_2AlH$
(DIBALH) at Low T

$LiAlH_4$

More reactive

We can predict the reactivity trend on the basis of charge, H-M bond polarity and steric bulk

Notes

- three ways to get H⁻

- will only be asked about $NaBH_4$ and $LiAlH_4$

- know that they result in an H⁻ nucleophile

- do not need to know how they form

Organolithium species (RLi) are a highly reactive source of anionic C:

$$R\text{—}X \;+\; Li \;\longrightarrow\; [R\text{—}X]^{\bullet-} \;+\; Li^{\oplus}$$

$$[R\text{—}X]^{\bullet-} \;\longrightarrow\; R^{\bullet} \;+\; X^{\ominus}$$

$$R^{\bullet} \;+\; {}^{\bullet}Li \;\longrightarrow\; R\text{—}Li$$

NET REACTION: $R\text{—}X \;+\; Li \;\longrightarrow\; R\text{—}Li \;+\; LiX$

This is an example of an:

(A) oxidative addition

One safety note: organolithium species are often **pyrophoric** (spontaneously ignite in air) and can be extremely dangerous to handle without specialized training.

Notes

• only need to know the net reaction

Lesson V.2. Preparation of RLi, Grignard and Gilman Reagents
Grignard Reagent preparation

One type of easily prepared organometallic species are the **Grignard reagents** (RMgX, where X = Cl, Br or I).

Preparation:

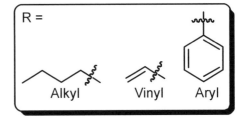

R =

Alkyl   Vinyl   Aryl

Oxidation states:

Mg

-1 +1   +1 -1
Cl—Mg—C

(A) *oxidative addition*

Grignard reagents are not quite as reactive as organolithium species, so they are (although still dangerous and requiring specialized training to handle) more desirable for lab use.

Notes

- think of organolithium and grignard reagents as "R⁻"

Lesson V.2. Preparation of RLi, Grignard and Gilman Reagents
Schlenk equilibrium

You may have noticed that we specify an ether solvent for preparation of Grignard reagents. In the absence of an ether, Grignard reagents actually exist as an equilibrium mixture of multiple species due to ligand exchange:

$$2 \ R{-}Mg{-}X \ \rightleftharpoons \ \text{(A)} \quad \boxed{}$$

This process for Grignard reagents specifically is called the **Schlenk equilibrium,** and it complicates the reactivity of Grignard reagents. In ether solvents, the O of the ether coordinates to the Mg center and stops this process:

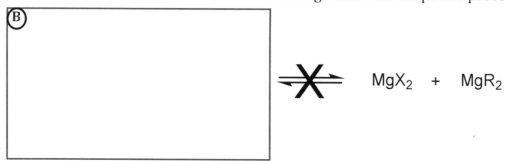

$$\text{(B)} \quad \boxed{} \ \xcancel{\rightleftharpoons} \ MgX_2 \ + \ MgR_2$$

Notes

Lesson V.2. Preparation of RLi, Grignard and Gilman Reagents
Basicity and nucleophilicity of Grignard reagents

Although Grignard reagents are good sources of nucleophilic C, they are also strongly basic. This causes problems in cases where acid-base and nucleophilic reactions compete (like E2 versus S_N2):

1. As a base:

$$H_3C^- + H \overset{..}{\underset{..}{O}} H \rightarrow H_3C-H + {}^-OH + heat$$

Grignard Reagents will also deprotonate more acidic groups, i.e.:

 (A)   you can't have any O-H or N-H around R^-!

2. As a nucleophile:

$H_3C^- = $ good nucleophile $H_3C-\overset{CH_3}{\underset{|}{C}}^--CH_3 = $ poor nucleophile

(B)   unless it is bulky, $R^- = $ good nucleophile

In cases where lower basicity is needed (to favor substitution, for example), scientists have developed less reactive organometallics of the form $LiCuR_2$, called **Gilman Reagents**.

Notes

Lesson V.2. Preparation of RLi, Grignard and Gilman Reagents
Preparation of Gilman reagents

Preparation from RLi:

2 〰〰Li + CuI ⟶ (A)

Li$^+$ $^-$Cu 〰〰

Reaction with RX:

(〰〰)$_2$CuLi + Br〰〰 ⟶ (B)

only substitution
no elimination

〰〰〰

Not a simple S$_N$2

because it works on
ternary sites

Notes

• Li$^+$ $^-$Cu 〰〰 , 〰〰CuLi, and LiCuBu$_2$ are all the same thing

Because the substitution induced by a Gilman reagent is not a simple S_N2 reaction, it works on both sp^3- and sp^2-hybridized carbons as well (recall that S_N2 only works on sp^3-hybridized C):

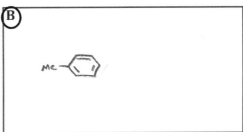

Me₂CuLi + [alkene with Br]

will not work for an S_N2 reaction

(A)

Me₂CuLi + [iodobenzene]

(B)

Notes

- one leaving groups one substitution
- only add one of the Rs

Lesson V.3. Reaction of Organometallics/MH with RX and Epoxides
Gilman reagents in substitution reactions

Me₂CuLi +

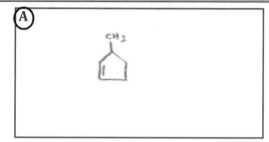

Me₂CuLi +

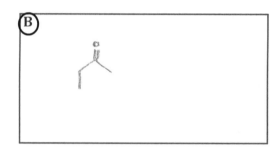

Notes

Epoxides can be used as starting materials to add additional carbon atoms if a carbanion source like RMgX or RLi is used:

Organometallics are strong bases, so:

Ⓒ

Nu attacks the less substituted side of the epoxide

Notes

• this reaction also works with hydrides

• carbonyls undergo nucleophilic addition with metal hydrides, RMgX, or RLi

R' from RLi or RMgx

• does not work when gilman reagent is used

Organopalladium compounds can undergo reversible oxidative addition with various organic compounds, such as aryl halides:

$$\text{Ph}_3\text{P—Pd—PPh}_3 \underset{-\text{ Ar–X}}{\overset{+\text{ Ar–X}}{\rightleftharpoons}}$$

Ⓐ

$$\underset{\text{Ar}}{\overset{\overset{X}{|}}{\text{Ph}_3\text{P} - \underset{|}{\text{Pd}} - \text{PPh}_3}}$$

The reversible nature of this process makes it a good candidate for catalytic CC bond-forming reaction. As we will see, many such reactions have been discovered and they have become so useful that the pioneers of the field won the Nobel Prize in Chemistry for this work.

Notes

• reacting palladium with an alkyl halide

• Ar = an aromatic group

A generalized Pd-catalyzed C-C bond forming reaction is given here:

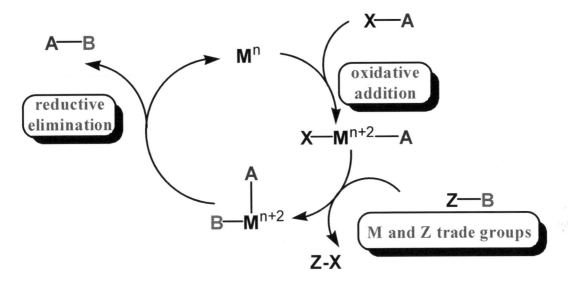

Notes

· M^n = palladium

· will not be asked specifics about the cycle

don't need to know!!

Lesson V.4. Palladium-Catalyzed C–C Bond-Forming Reactions
Kumada reaction

The **Kumada Reaction** couples:

(A)

R = aryl, benzyl or vinyl; X = OTf, Cl, Br, or I

A simplified catalytic cycle:

An example:

Ar–R ← Ph₃P—Pd—PPh₃ → Ar–X

step 3   step 1

PPh₃
Ph₃P—Pd—Ar
R

PPh₃
Ph₃P—Pd—Ar
X

MgX₂ ← step 2 ← R–MgX

PhMgCl + 4-iodotoluene

↓ Pd(PPh₃)₄

(B)

<u>*Notes*</u>

118

Lesson V.4. Palladium-Catalyzed C–C Bond-Forming Reactions
Heck reaction

The **Heck Reaction** couples:

(A)

aromatic group and an alkene

R = aryl, benzyl or vinyl; X = OTf, Cl, Br, or I

$$R-X \; + \; H \underset{H}{\overset{H}{=}} R' \quad \xrightarrow[N(C_2H_5)_3]{Pd(PPh_3)_4} \quad R \overset{R'}{=}$$

☆ is on exam !

(B)

$$H_3CO- \bigcirc -OTf \; + \; \underset{H}{\overset{H}{=}} \bigcirc \quad \xrightarrow[N(C_2H_5)_3]{Pd(PPh_3)_4}$$

(C)

$$\underset{I}{\overset{O}{\bigcirc}}-OCH_3 \; + \; \underset{H}{\overset{H}{=}}\underset{H}{\overset{H}{}} \quad \xrightarrow[N(C_2H_5)_3]{Pd(PPh_3)_4}$$

Notes

119

A simplified mechanism for the Heck Reaction:

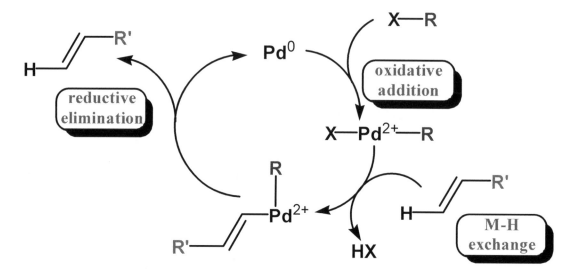

Notes

Lesson V.4. Palladium-Catalyzed C–C Bond-Forming Reactions
Stille Reaction

The **Stille Reaction** couples:

aromatic group with either an alkene or another aromatic group

R = aryl, benzyl or vinyl; X = OTf, Cl, Br, or I

$$R-X + R'-SnBu_3 \xrightarrow[THF]{Pd(PPh_3)_4} R-R'$$

★ is on exam 1

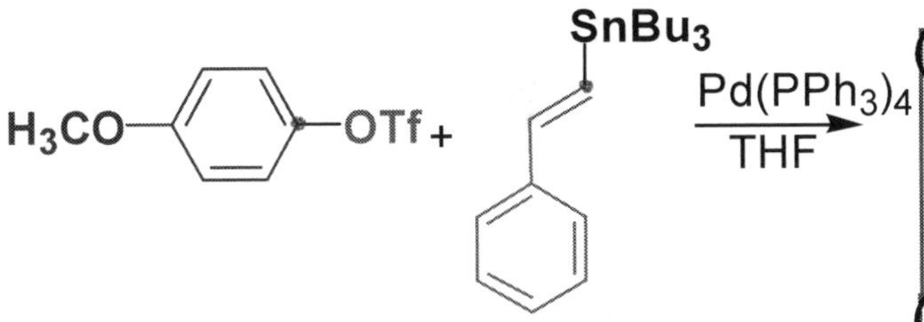

Bu: butyl

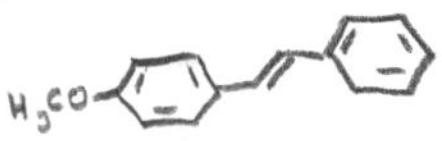

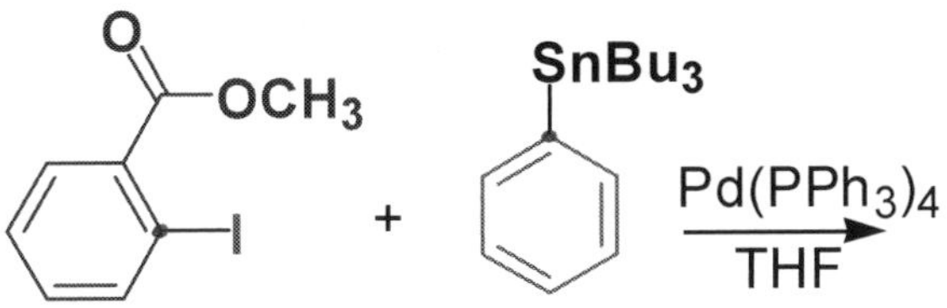

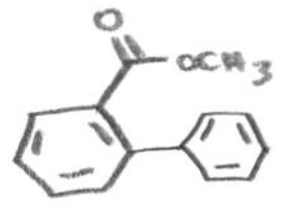

Notes

A simplified mechanism for the Stille reaction is:

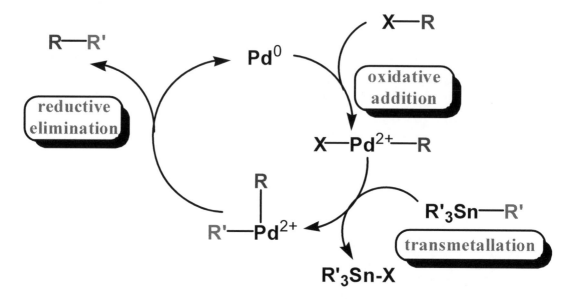

<u>*Notes*</u>

Lesson V.4. Palladium-Catalyzed C–C Bond-Forming Reactions
Sonogashira reaction

The **Sonogashira Reaction** couples:

(A)

R = aryl, benzyl or vinyl; X = OTf, Cl, Br, or I

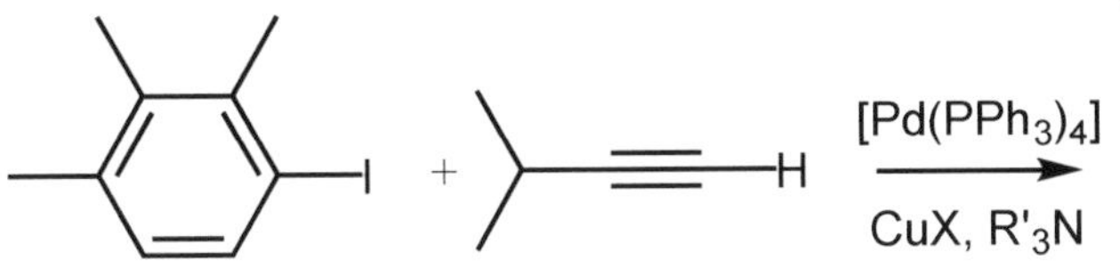

I
+
H
[Pd(PPh3)4]
CuX, R'3N

(B)

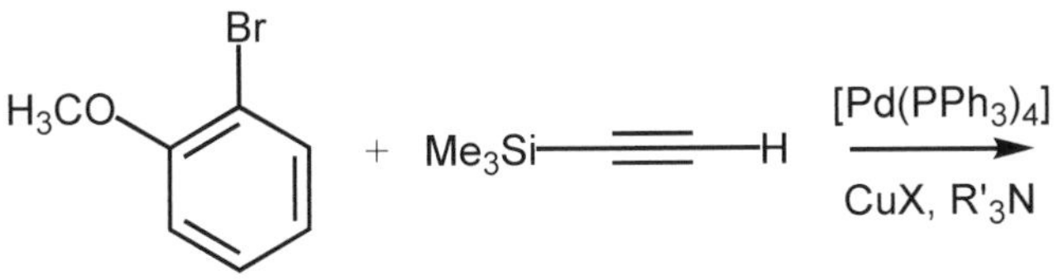

Br
H3CO
+
Me3Si
H
[Pd(PPh3)4]
CuX, R'3N

(C)

Notes

A simplified mechanism for the Sonogashira reaction is:

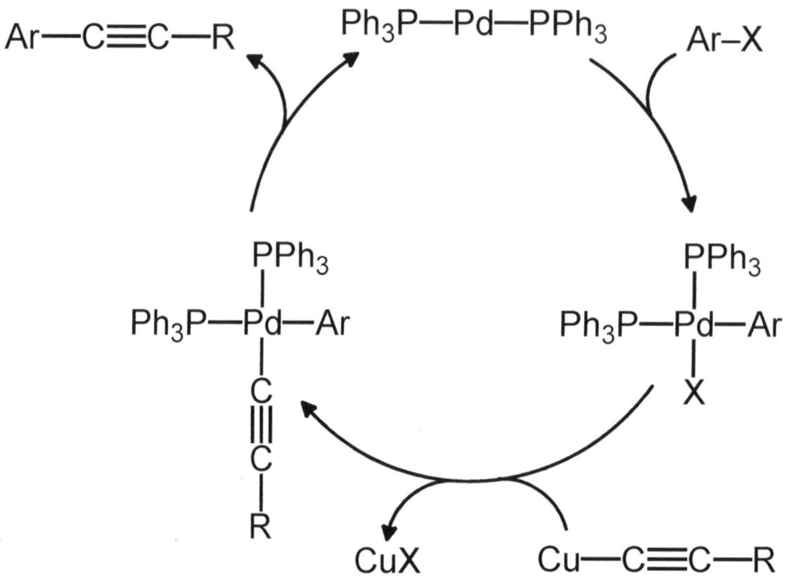

Notes

The **Suzuki Reaction** couples:

(A)

two aromatic groups or an aromatic group and an alkene

R = aryl, benzyl or vinyl; X = Cl, Br, or I

$$R\text{--}X + R'\text{--}B(OR)_2 \xrightarrow[\text{NaOH}]{\text{Pd(PPh}_3)_4} R\text{--}R'$$

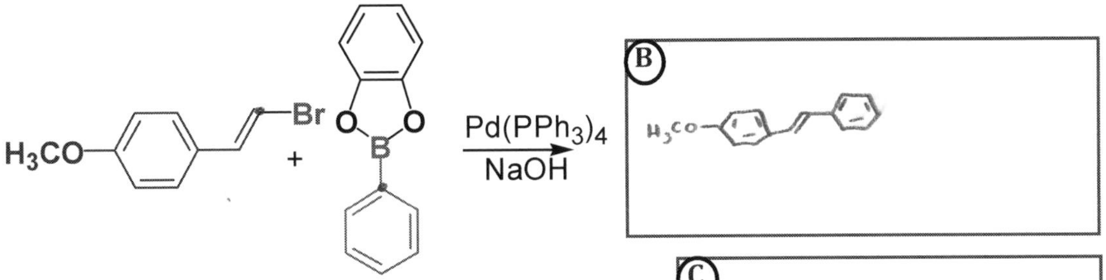

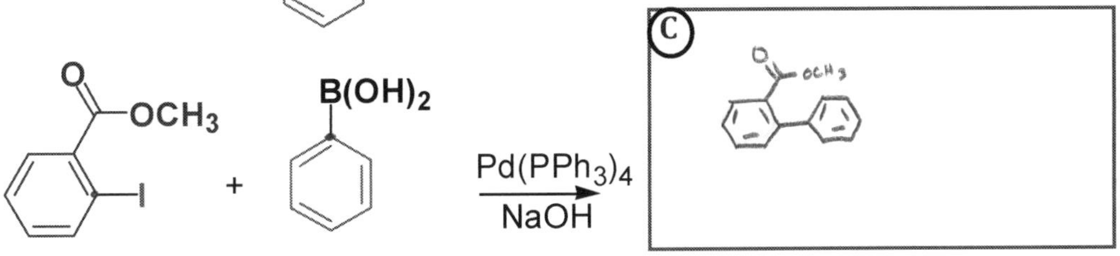

(B)

(C)

Notes

A simplified mechanism for the Suzuki reaction is:

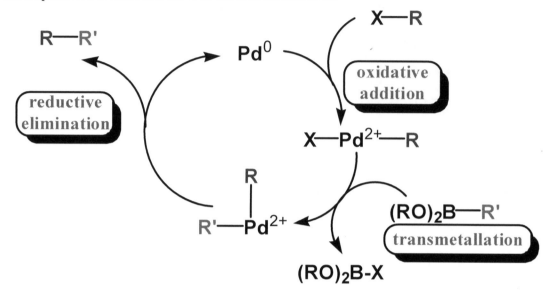

Notes

Prof. Richard Schrock (now at MIT) worked with early transition metals and noted that some metals having a C=M bond could cause "scrambling" of alkenes:

This reaction involves *breaking* the C=C double bond and *rearranging* the two doubly-bound units! This is a process called **alkene metathesis**. This reaction has become so important that three of the pioneers of the field won the Nobel Prize in Chemistry for this work.

<u>*Notes*</u>

Yves Chauvin did some work to elucidate the mechanism my which a simpler metathesis reaction takes place:

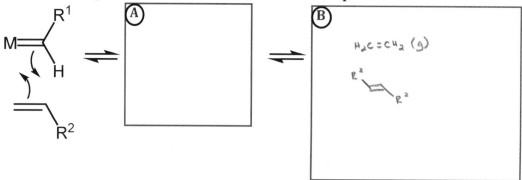

The major product is generally the more stable of the possible alkenes.

<u>*Notes*</u>

Robert Grubbs developed catalysts that are stable in the air (this is a big improvement over many of the air-sensitive organometallic reagents we have talked about). These catalysts use ruthenium (Ru) as the metal:

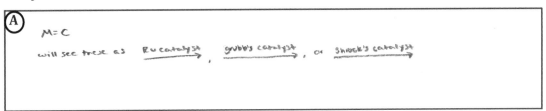

first ruthenium alkylidene catalyst   **first-generation Grubbs catalyst**   **second-generation Grubbs catalyst**   *Schrock*

Key features:

(A)

M = C

will see these as   Ru catalyst ⟶ ,   grubb's catalyst ⟶ ,   or   Shrock's catalyst ⟶

Notes

• do not need to know the structure of these.

The Grubbs catalysts can be used to mediate many useful reactions, including Cross metathesis. This reaction is driven to one major product by removal of ethylene gas (ethenolysis).

Example:

Catalytic Cycle:

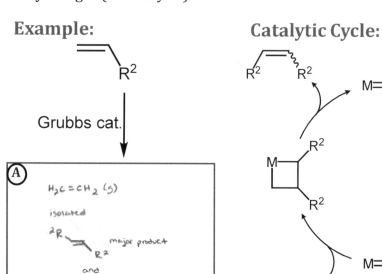

$H_2C = CH_2$ (g)

isolated

major product

and

minor product

gas removed
during reaction

Notes

Lesson V.6. Applications of Alkene Metathesis
Ring-closing metathesis (RCM)

For favorable ring sizes (5- to 7-carbons), Grubbs catalyst can be used for Ring-Closing Metathesis (RCM):

General:

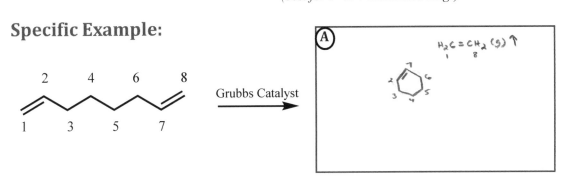

flexible alkyl chain

flexible alkyl chain

Grubbs Catalyst

$H_2C=CH_2$

an α,ω-diene

*more effective for more
favorable ring sizes
(best for 5- to 7-membered rings)*

Specific Example:

2   4   6   8

1   3   5   7

Grubbs Catalyst

(A)

$H_2C=CH_2$ (g) ↑

Notes

On the other hand, *strained rings* can undergo **Ring-Opening Metathesis Polymerization (ROMP)** with Grubbs catalyst:

General:

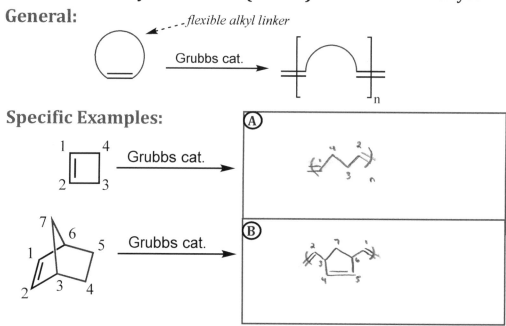

Specific Examples:

Notes

PART VI. Properties and Reactivity Trends of Carbonyl Compounds

Lesson VI.1. Nomenclature of Carbonyls

Prioritizing functional groups

Hierarchy of functional group priorities for naming:

Carboxylic Acid

(A)

Acid Chloride

(B)

Aldehyde

(C)
ketone

Alcohol

Notes

Lesson VI.1. Nomenclature of Carbonyls
Naming Ketones

Ketones: 1) Use the *–one* ending; 2) number to indicate which C has the =O on it:

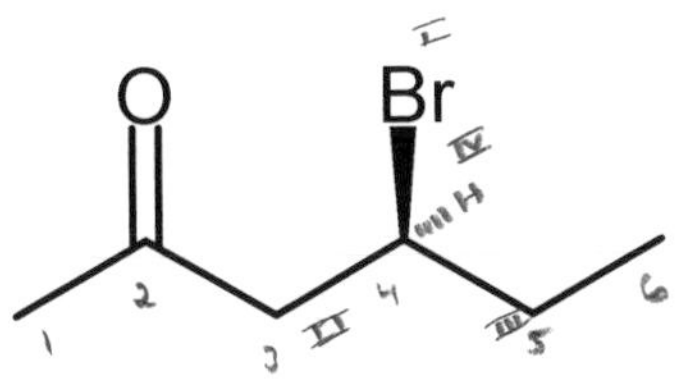

(A) (S)-4-bromo-2-hexanone

A ketone has priority over an alcohol. Name the alcohol as a *"hydroxy"* substituent:

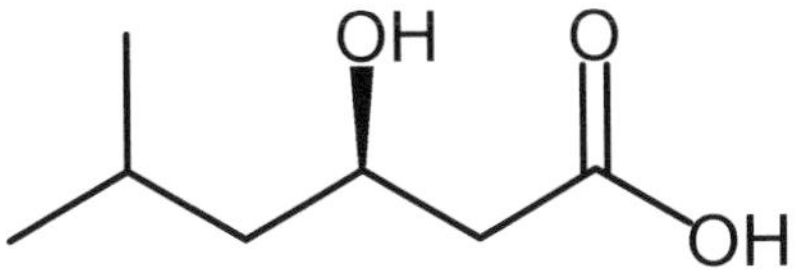

(B)

Notes

Aldehydes: 1) Use the *–al* ending; 2) it is always at the end; 3) as a substituent, name it "carbaldehyde" or "formyl"

4-ethyl-3,5-dimethylhexanal

cyclopentanecarbaldehyde
or formylcyclopentane

An aldehyde has priority over an alcohol or a ketone. Name the alcohol as a *"hydroxy"* substituent and ketone as a *"keto"* or *"oxo"* substituent:

(A) 2-hydroxy-4-ketopentanol

hydroxy   keto

Notes

Lesson VI.1. Nomenclature of Carbonyls
Naming acid chlorides

Acid chlorides: 1) Use the "*–oyl chloride*" ending; 2) the acid chloride is always at the end, so it does not need a number to indicate its position.

Acid Chloride has priority over aldehyde, ketone and alcohol functional groups:

Ⓑ

Notes

Esters: 1) place the name of the substituent on the non-carbonyl O in front of the name; 2) use the "*-oate*" ending. 3) the ester is always on the end, so the name will not need a number indicating its position.

fig

(A)

A carboxylic acid has priority over all of the other carbonyl functional groups other than carboxylic acid that are covered in this Lesson.

(B)

Notes

Lesson VI.1. Nomenclature of Carbonyls

Naming carboxylic acids

Carboxylic acids: 1) use "-oic acid" in place; 2) it is always on the end

SKIP

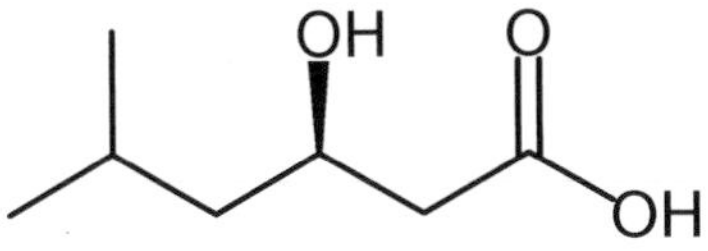

(A)

A carboxylic acid has priority over all of the other carbonyl functional groups in this Lecture Guide.

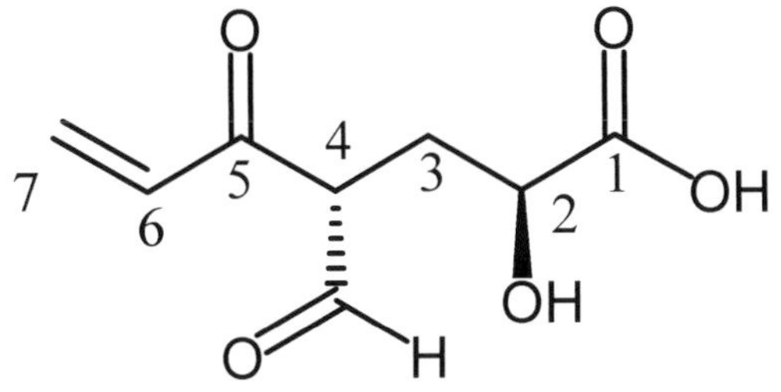

(2*S*, 4*R*)-4-formyl-2-hydroxy5-ketohept-6-enoic acid

Notes

We have seen several ways to make carbonyl-containing functional groups. One way is to oxidize an alcohol:

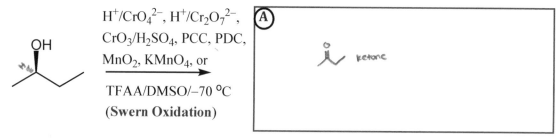

PCC = pyridinium chlorochromate, PDC = pyridinium dichromate, TFAA = trifluoroacetic anhydride, $CF_3C(O))_2O$.

The secondary alcohols are oxidized to ketones by all of these reagents.

<u>*Notes*</u>

Lesson VI.2. Review: Preparation of Carbonyls from Alcohols, Alkynes and Alkenes

Preparation of carbonyls from alcohols

Alcohols or benzylic alkyl group can be oxidized:

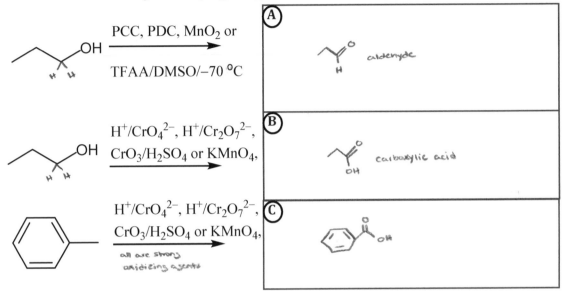

Primary alcohols can be used as precursors for aldehydes **or** carboxylic acids, depending on the oxidizing agents used.

Notes

Carbonyl functional groups can be made from alkynes by oxymercuration/demercuration or hydroboration/oxidation:

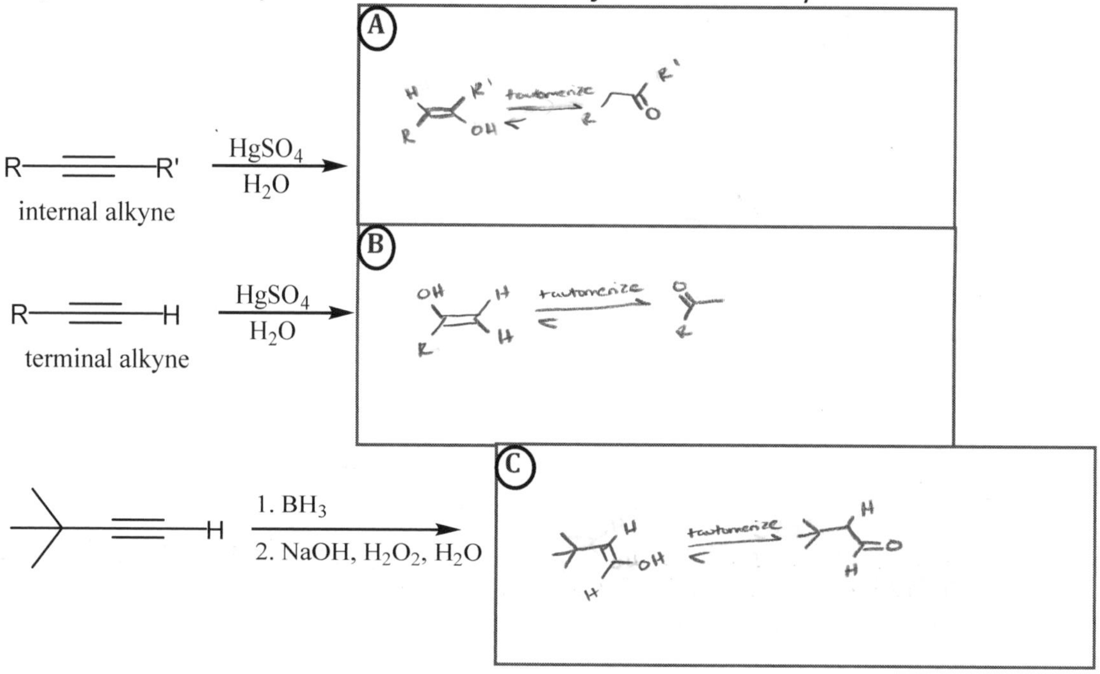

Notes

Preparation of carbonyls from alkenes

Finally, we can use ozonolysis of an alkene to produce aldehydes, ketones or carboxylic acids:

Note that whether the sp^3-carbon-bound H remains or is oxidized depends on the workup conditions (step 2).

Notes

Type A: Single Nucleophilic Addition

The net result is:

1. Add a nucleophile (Nu) to the carbonyl C to replace the pi bond

2. Protonate the carbonyl O.

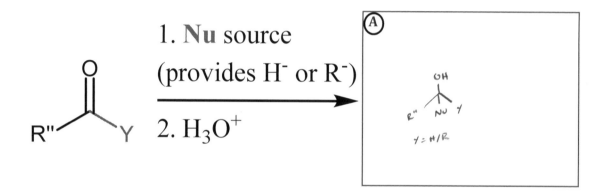

Notes

Type B: Nucleophilic Acyl Substitution (S$_N$Ac)

The net result is:

1. Substitute one nucleophile (Nu) for one leaving group (Y) attached to the carbonyl carbon

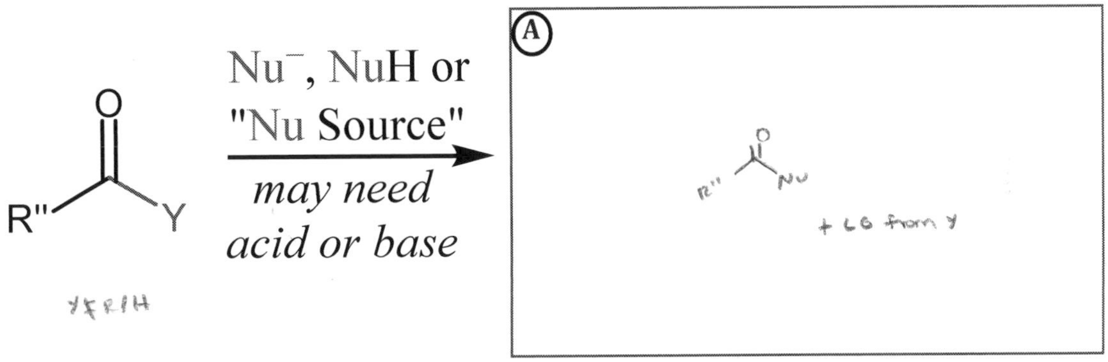

Nu⁻, NuH or "Nu Source"
may need acid or base

Ⓐ

+ LG from Y

<u>Notes</u>

Type C: S$_N$Ac then Nucleophilic Addition

(Type C = B then A!)

The net result is:

1. Replace the pi bond to O and the leaving group with two bonds to nucleophiles.

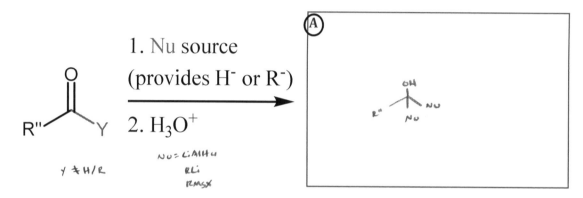

1. Nu source

(provides H⁻ or R⁻)

2. H$_3$O⁺

Notes

Type D: Replace both Bonds to the Carbonyl O

The net result is:

1. Remove the carbonyl O.

2. Replace the two bonds to carbonyl C. There are four options to replace the two bonds ...

$$\overset{O}{\underset{}{\lambda}}{}_1 \to X_1$$

Notes

147

Type D: Replace both Bonds to the Carbonyl O

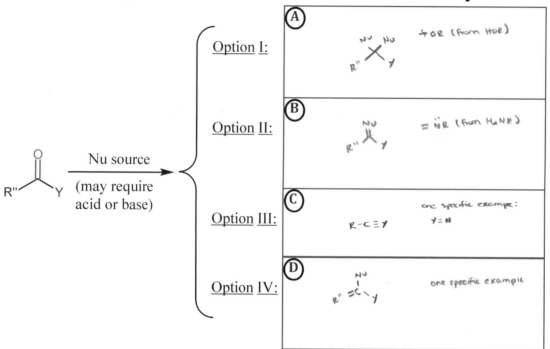

Option I:

Option II:

Option III:

Option IV:

_____

Notes

Lesson VI.4. Relative Rates of Nucleophilic Attack on Carbonyl Functional Groups

Carbonyl reactivity depends on sterics and electronics

Because each of the various carbonyl-containing functional groups has a different substituent, their reactivities to the initial nucleophilic addition vary as well:

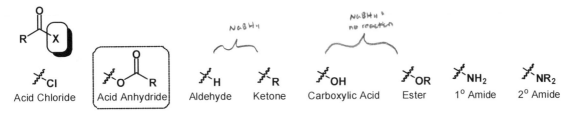

| Acid Chloride | Acid Anhydride | Aldehyde | Ketone | Carboxylic Acid | Ester | 1° Amide | 2° Amide |

Less electron density (more positive charge) on the carbonyl C will increase susceptibility to nucleophilic attack:

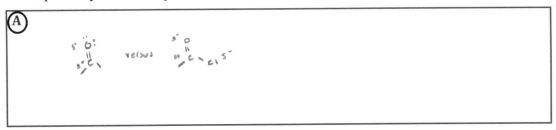

Notes

Lesson VI.4. Relative Rates of Nucleophilic Attack on Carbonyl Functional Groups

Carbonyl reactivity depends on sterics and electronics

Drawing out the contributing resonance structures can help us assess the electron-density at the carbonyl unit:

more negative on O, so a reaction pushing - at O is slow a

| Resonance contributors | Resonance hybrid |

Esters will have similar resonance structures wherein an O lone pair moves

Acid (or metal coordination) can increase the reactivity of a carbonyl:

Notes

Lesson VI.4. Relative Rates of Nucleophilic Attack on Carbonyl Functional Groups
Carbonyl reactivity depends on sterics and electronics

More steric encumbrance slows down the rate of nucleophilic attack (just like we saw for S_N2 reactions, 1° > 2° > 3°):

Aldehyde versus ketone:

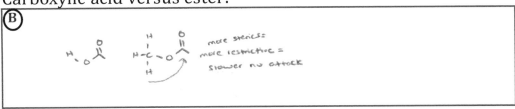

Carboxylic acid versus ester:

A similar case can be made for comparison of primary and secondary amides.

Notes

Carbonyls can undergo **nucleophilic addition** with metal hydrides, RMgX or RLi:

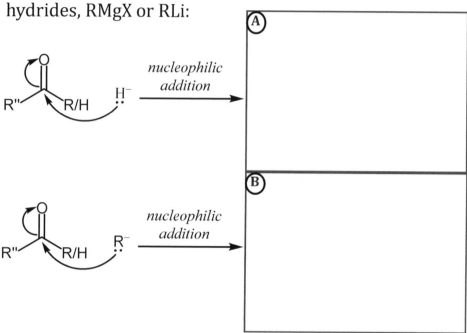

Ⓐ

Ⓑ

Notes

Aldehydes and Ketones do the same reactions as one another throughout the course!

Metal hydride reactions with ketones and aldehydes

The metal hydrides will do Type A reactions with aldehydes and ketones to give alcohols. These are **reduction** reactions.

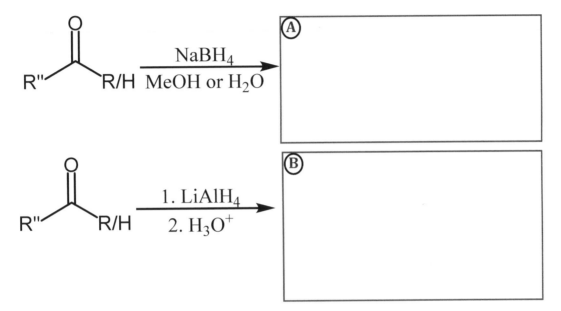

A

B

Notes

The Grignard reagents and organolithium reagents will do Type A reactions with aldehydes and ketones. Gilman reagents do **not** do nucleophilic addition to ketones or aldehydes.

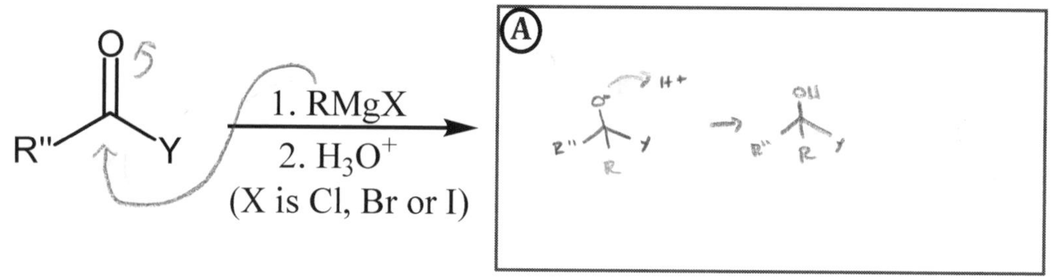

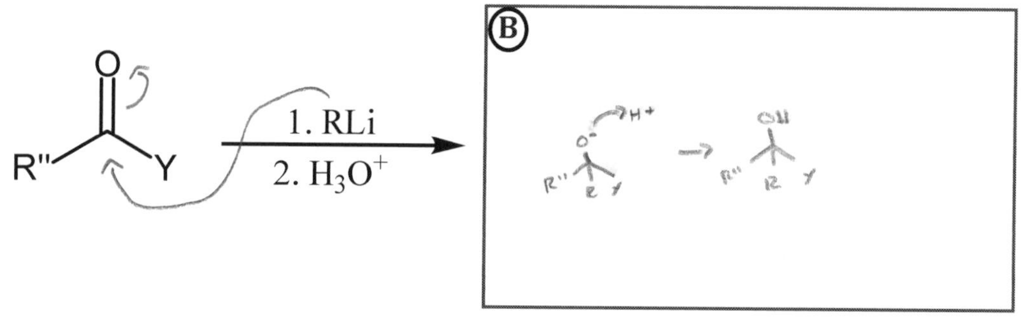

(A)

1. RMgX
2. H_3O^+
(X is Cl, Br or I)

(B)

1. RLi
2. H_3O^+

Notes

Lesson VI.6. Addition of Water or Alcohols to Aldehydes and Ketones

Water reacts with aldehyde/ketone to form hydrates

Neutral nucleophiles like water and alcohols are also capable of doing Type A reactions with aldehydes and ketones. When water is the nucleophile, a **hydrate** is formed:

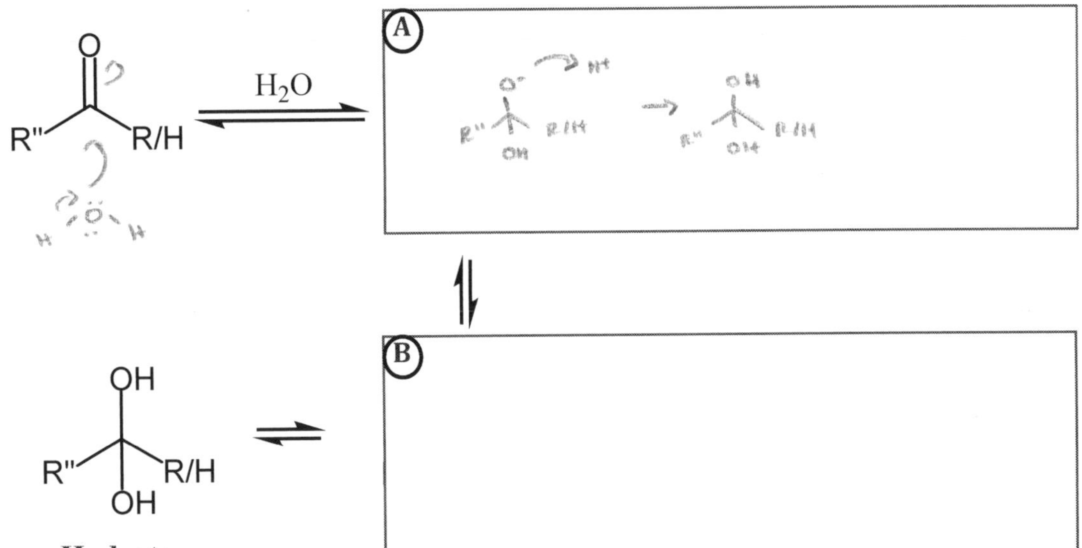

Hydrate

Notes

When an alcohol is the nucleophile, a **hemiacetal** is formed from an aldehyde and a **hemiketal** is formed from a ketone:

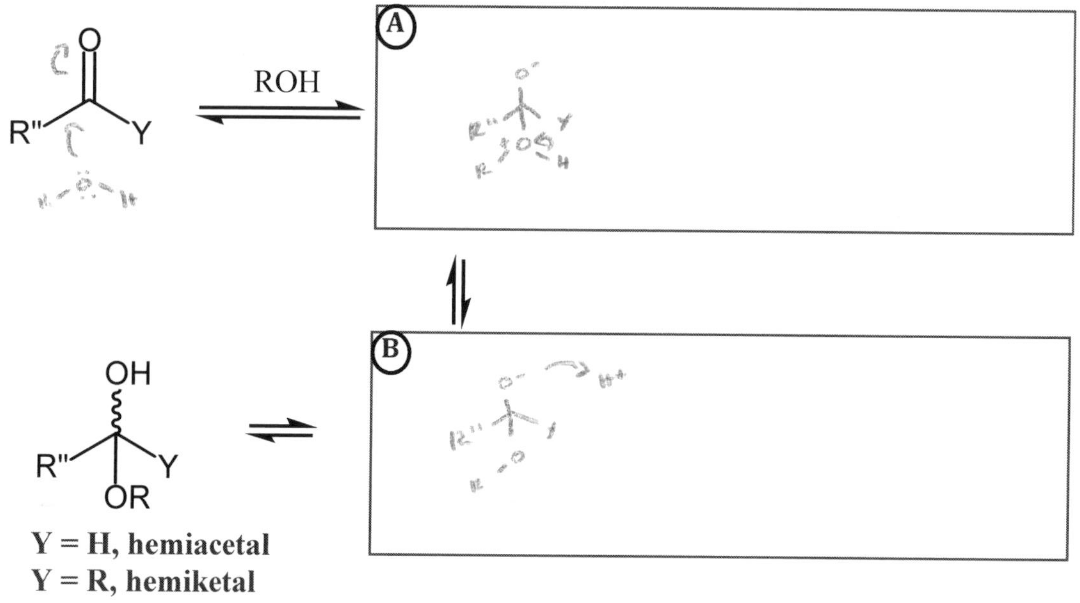

ROH

(A)

(B)

Y = H, hemiacetal
Y = R, hemiketal

Notes

We have already seen that aldehydes and ketones can react with ROH to form hemiacetals and hemiketals, respectively. If we continue reaction, aldehydes and ketones will form **acetals** and **ketals**, respectively:

Y=H: hemiacetal
Y=R': hemiketal

−H₂O

Y=H: **Acetal**
Y=R': **Ketal**

Notes

This is a Type D reaction (option I):

The carbonyl double bond to O is replaced by two bonds to OR!

Acetals/ketals are treated in some detail by your text because of their importance as **protecting groups:**

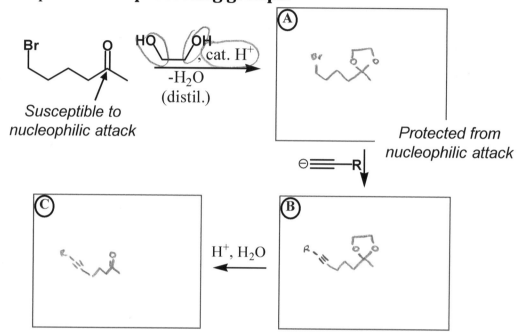

Susceptible to nucleophilic attack

$-H_2O$
(distil.)

Ⓐ

Protected from nucleophilic attack

H^+, H_2O

Ⓒ

Ⓑ

Notes

When we react a ketone with a 1° amine, we form an **imine (Schiff base)**:

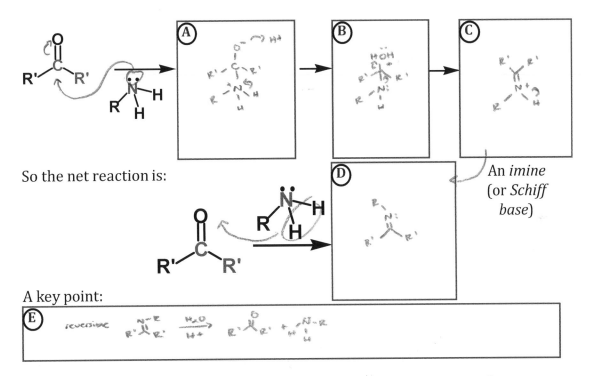

So the net reaction is:

An *imine*
(or *Schiff
base*)

A key point:

Notes

This is a Type D reaction (option II):

The carbonyl double bond to O is replaced by two bonds to NR!

Secondary amines lack the second proton that has to leave in order to form the neutral imine, so if we react a ketone or aldehyde with a secondary amine, the initial stages of the reaction are similar to those observed on the previous slide. However, when it comes time for the second proton to leave, a different proton needs to be used:

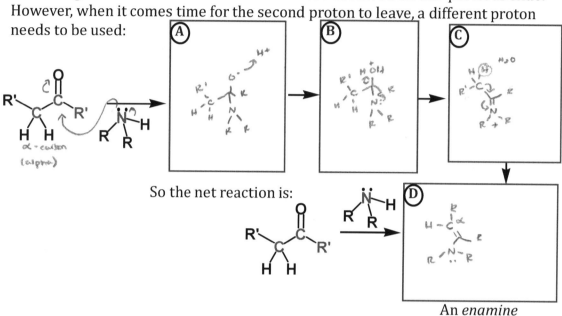

So the net reaction is:

An *enamine*

Notes

This is a Type D reaction (option 4):

The carbonyl double bond to O is replaced by one bond to NR$_2$ and one bond to the α-carbon!

Two other ways to reduce ketones all the way to alkyl groups:

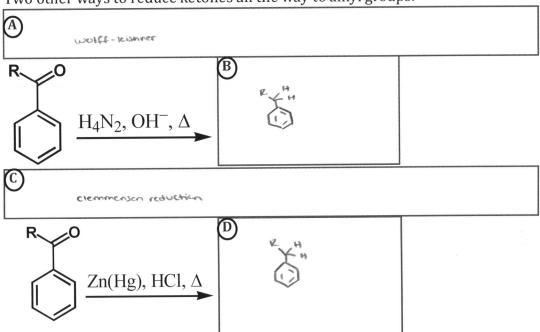

Ⓐ wolff-kishner

Ⓑ

$$H_4N_2, OH^-, \Delta$$

Ⓒ clemmensen reduction

Ⓓ

$$Zn(Hg), HCl, \Delta$$

** these work on ketones that are NOT next to benzene as well (unlike H_2/Pd)

Notes

This is a Type D reaction (option I):
The carbonyl double bond to O is replaced by two bonds to H!

A special type of nucleophile called a **phosphonium ylide** can be formed from a phosphonium salt, and then used as a nucleophile to do a Type D reaction with aldehydes and ketones. This particular Type D reaction is called the **Wittig Reaction**, and produces alkene products.

Here is how the phosphonium salt and the phosphonium ylide are generated in the Wittig reactions:

triphenylphosphonium halide

triphenylphosphonium ylide

(from *n*-BuLi)

S_N2

−butane

Notes

This is a Type D reaction (option II):

The carbonyl double bond to O is replaced by two bonds to CR_2!

Lesson VI.8: The Wittig Reaction
Making an alkene from a carbonyl!

Once the ylide is generated, it reacts with the carbonyl by a process similar to the Chauvin Mechanism we saw for reaction of a metal alkylidene with an alkene:

Alkene

One of the key driving forces for this reaction is the formation of a P=O bond in the $Ph_3P=O$ product; the P–O bonds are among the strongest covalent bonds.

Note that alkenes can exhibit *E-/Z-* and *cis-/trans-* isomers. The major alkene product formed is:

Z- alkene

Notes

One type of reaction that carbonyl functional groups can undergo consists of two steps: *1) nucleophilic addition then 2) nucleophilic elimination.* These steps comprise **Nucleophilic Acyl Substitution (S$_N$Ac).** In this Lecture Guide, we also refer to these type of reactions as **carbonyl reaction type B:**

Type B reactions are possible for all of the carbonyl functional groups covered in this Lecture Guide, **except** aldehydes and ketones.

Notes

Lesson VI.9. Nucleophilic Acyl Substitution of Acid Chlorides and Anhydrides
Acid chlorides readily undergo SNAc

The situation is much simpler when nucleophiles react with acid chlorides or anhydrides. In these cases, there is already a good leaving group on the carbonyl carbon. A variety of nucleophiles react readily with these functional groups. Consider acid chlorides:

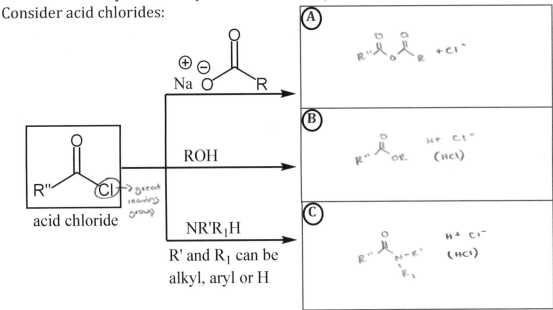

acid chloride

Lesson VI.9. Nucleophilic Acyl Substitution of Acid Chlorides and Anhydrides
Anhydrides readily undergo S$_N$Ac

And anhydrides:

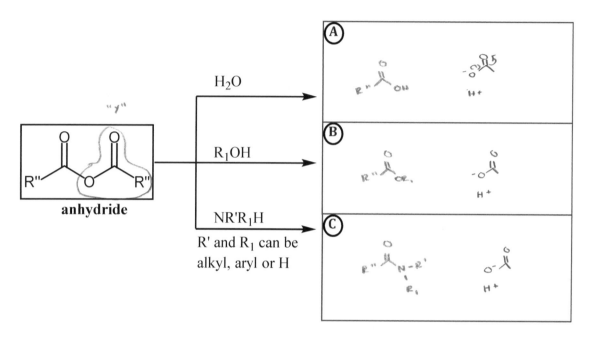

$$\boxed{\begin{array}{c} \text{anhydride} \end{array}}$$

Notes

Carboxylic acids are not very good substrates for S$_N$Ar. This is because the –OH group is a bad leaving group. For alcohols, we converted the –OH into a good leaving group *before* we did a substitution by using Thionyl chloride (SOCl$_2$) or phosphorus trihalide (PX$_3$, X = Cl, Br).

These reagents also work for carboxylic acids:

Notes

Lesson VI.10. S$_N$Ac of Carboxylic Acids
Activation of carboxylic acids

Here is the reaction mechanism for a carboxylic acid with thionyl chloride:

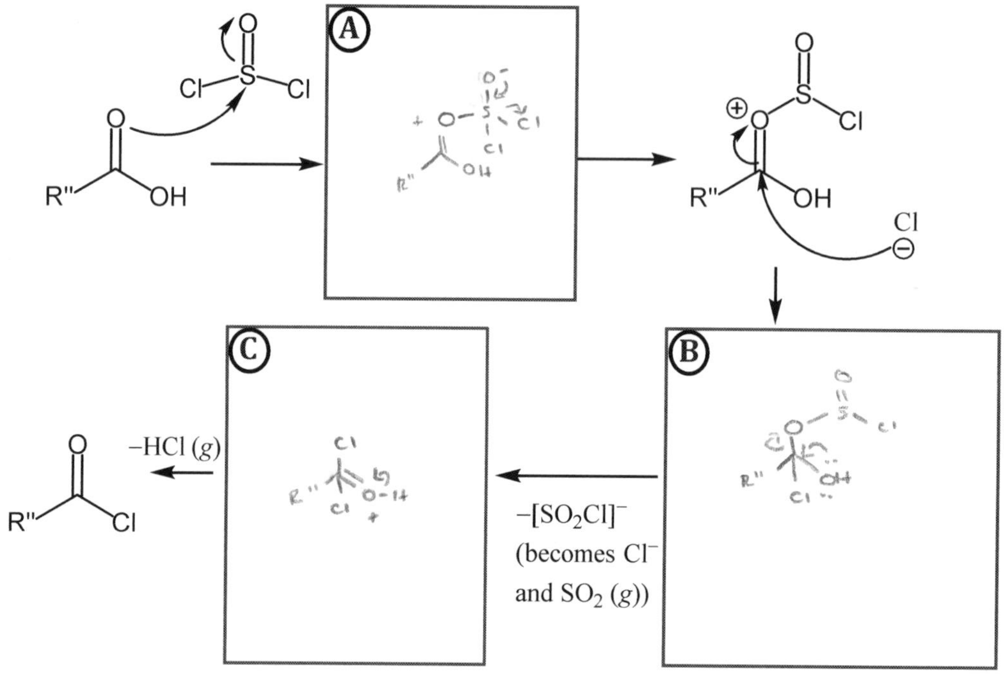

−HCl (g)

−[SO$_2$Cl]$^-$
(becomes Cl$^-$
and SO$_2$ (g))

Notes

Lesson VI.10. SNAc of Carboxylic Acids

Activation of carboxylic acids

Here is the reaction mechanism for a carboxylic acid with phosphorus trichloride:

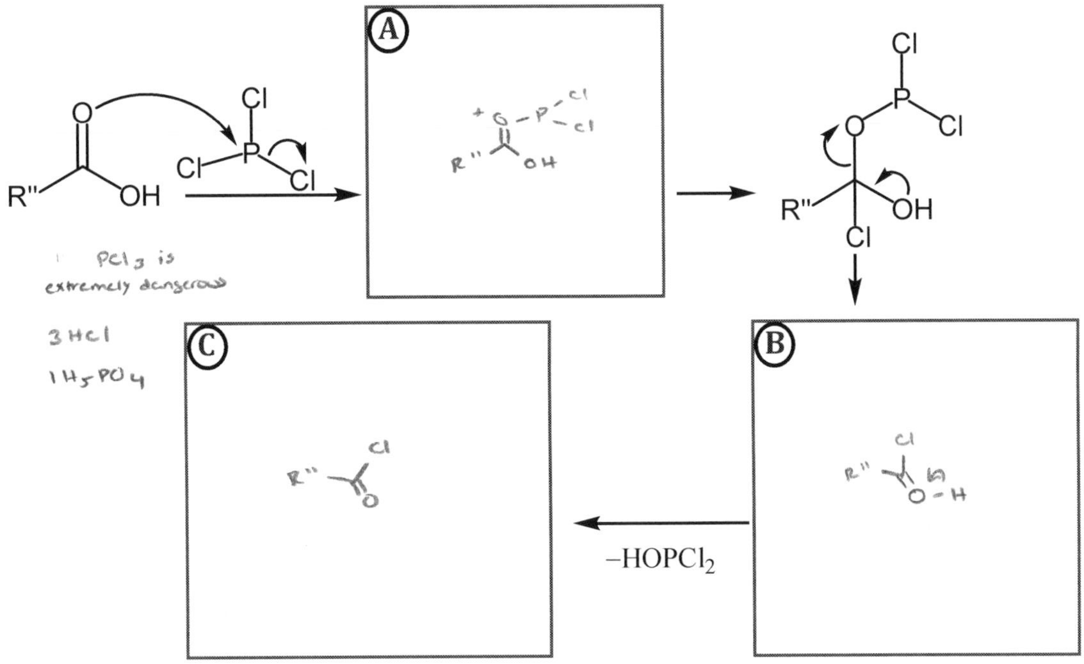
A
B
C
Cl
P
O
R''
OH
–HOPCl2
PCl3 is extremely dangerous
3HCl
1H3PO4

Notes

It is possible to convert an –OH group to a good leaving group by protonating it as well, as we saw in reactions of alcohols with HX or H_2SO_4. If we react a carboxylic acid with an alcohol nucleophile in the presence of an acid catalyst, we will get an ester. This is called **acid-catalyzed esterification** or **Fischer esterification**:

Notes

Note that all the steps are reversible, so we can push the reaction to either side of the equation by using LeChatelier's Principle.

LeChatelier's Principle: Ⓑ

> push equilibrium either way be either adding excess reagent or removing the product

This means that ester hydrolysis to convert an ester to a carboxylic acid is also possible:

If you add lots of water:

$+ xs\ H_2O$ NU $\xrightarrow{\text{cat. HCl}}$

Ⓒ

Carboxylic acid Favored

If you add lots of alcohol:

$+ H_2O$ $\xrightarrow{xs\ ROH}$

Ⓓ

Ester Favored

Notes

Lesson VI.11. S_NAc Reaction of O Nucleophiles with Carboxylic Acids and Esters
Base-catalyzed ester hydrolysis, transesterification

We have seen that esters can undergo acid-catalyzed hydrolysis. Hydrolysis can also be mediated by base catalysis. The presence of base leads to formation of a carboxylate as the product, so the final step is irreversible:

If an alcohol/alkoxide is used in place of the H_2O/hydroxide, one ester can be changed into a different ester, a process called **transesterification:**

Notes

Lesson VI.11. SNAc Reaction of Oxygen Nucleophiles with Carboxylic Acids and Esters
Acid-catalyzed transesterification

We have seen that esters can undergo acid-catalyzed hydrolysis. If an alcohol is used in place of water for reaction with ether with acid catalysis, **acid-catalyzed transesterification** is possible:

Note that this mechanism mirrors the Fischer esterification of carboxylic acids.

Notes

Ammonia, primary amines, or secondary amines can also be used as nucleophiles for S_NAc of esters:

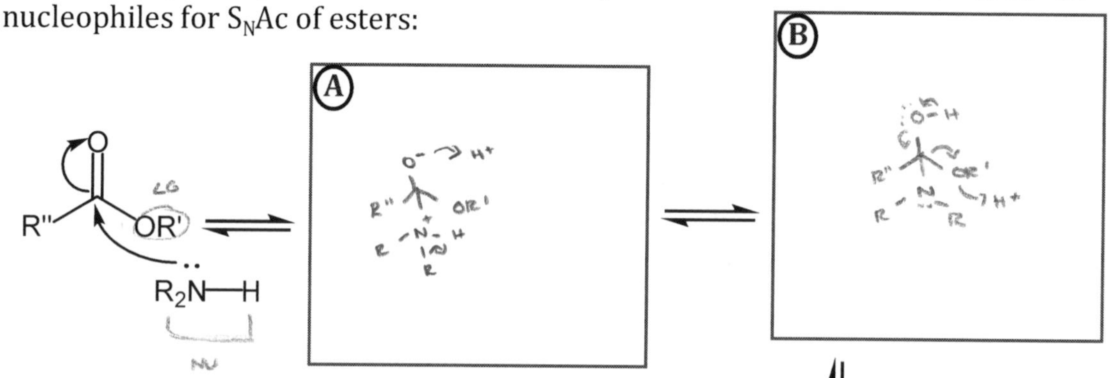

When ammonia (R= H) is used as the nucleophile, this reaction is called **ammonolysis**. When a 1° or 2° amine is used, this reaction is called **amidation**.

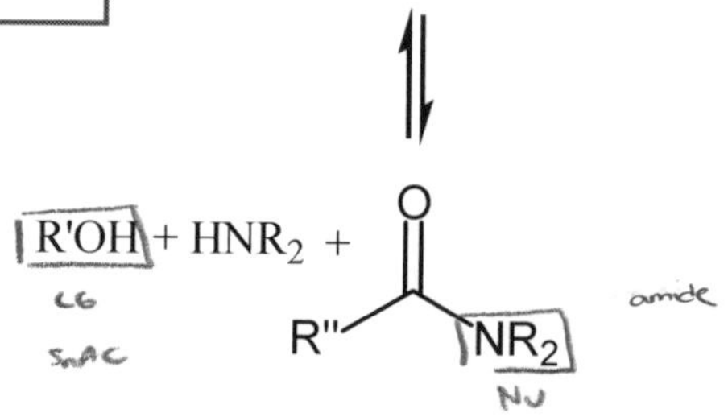

$$R'OH + HNR_2 +$$

Notes

Like esters, amides can undergo hydrolysis to form carboxylic acids or carboxylates. The amides are much less reactive to the initial nucleophilic addition step, so these reactions require higher temperatures and/or longer reaction times than do ether hydrolysis reactions, but mechanistically it is quite similar:

(R is alkyl, or aryl or H)

Nu attacks

LG leaves

Notes

Preparation of primary amines from alkyl bromides is very useful; however, if one uses ammonia as the nucleophile, up to four R groups can add to N:

mix of products

The Gabriel Synthesis yields 1° amines from **phthalimide** over a few steps:

1. NaOH
2. Primary RX
3. H_3O^+, H_2O
4. base

Phthalimide

What is the mechanism for the Gabriel Synthesis?

Notes

Lesson VI.12. Amide Formation, Amide Hydrolysis and the Gabriel Synthesis

Gabriel Synthesis

The Gabriel Synthesis yields 1° amines by protecting two sites of the N for the S_N2 reaction, then deprotection by acid hydrolysis:

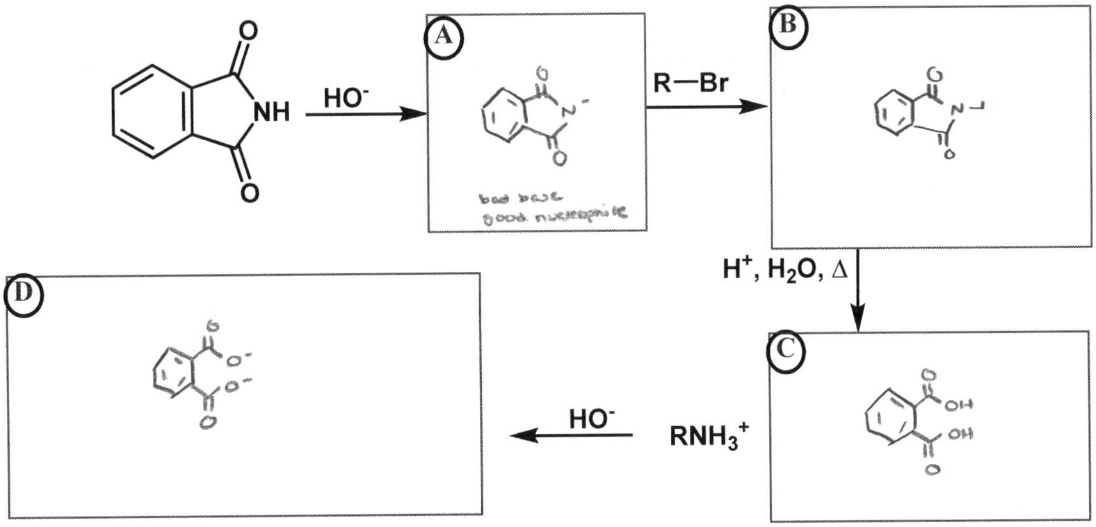
O
NH
O
HO-
A
bad base
good nucleophile
R—Br
B
H+, H2O, Δ
C
OH
OH
RNH3+
HO-
D

Notes

Recall when we examined the hydride sources to be used in this chapter, we noted that DIBALH is the least reactive of the hydride sources under discussion. In fact, if we use DIBALH as the hydride source and keep the temperature low (typically $-78\,°C$, we are able to stop the reduction of an ester at the aldehyde:

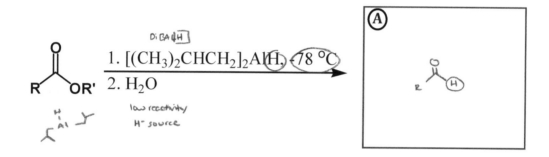

This is an important reaction for preparing aldehydes.

Notes

Some carbonyls undergo reaction that is essentially reaction Type B (S_NAc) followed by reaction Type A (nucleophilic addition then protonation). In this Lecture Guide, we will refer to these reactions as **Type C** for the purpose of sorting and discussion:

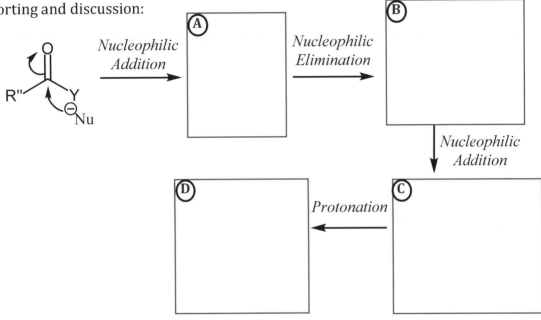

Lesson VI.13. Reaction of Carboxylic Acid Derivatives with H or C Nucleophiles
Metal hydrides with carboxylic acids, esters and acid chlorides

Type C reactions occur between $LiAlH_4$ and either carboxylic acids, esters, or acid chlorides to give 1° alcohols. $NaBH_4$ can also facilitate this reaction with acid chlorides, **but not with any other functional groups.**

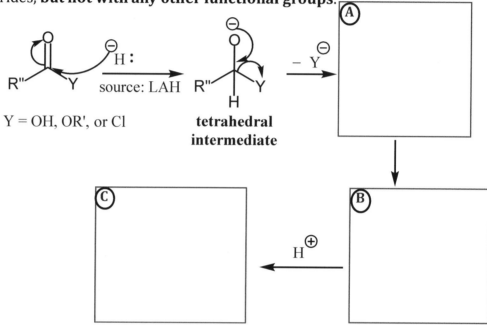

Y = OH, OR', or Cl

tetrahedral intermediate

Notes

Type C reactions occur between LiR or RMgX and either esters, or acid chlorides to give alcohols as well. LiR can also facilitate this reaction with carboxylic acids, but Grignard reagents will only deprotonating carboxylic acids to make the carboxylate. The mechanism is quite similar to the reaction with hydride:

Y = OR', or Cl

tertiary alcohol product
(secondary if R" is H)

acid work-up

Notes

181

Amides react with LiAlH$_4$ to give amines:

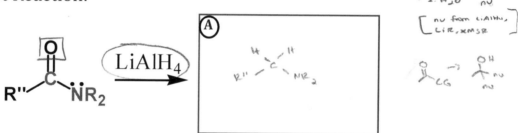

R is H, alkyl or aryl groups

Net Reaction:

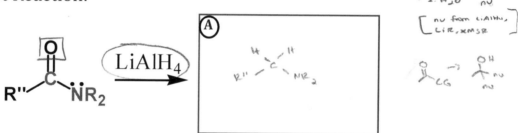

Notes

Amides can also be dehydrated by reaction with P_2O_5 to give **nitriles:**

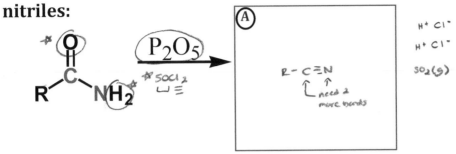

We will not go into the detailed mechanism, but it is worth noting that the nitrile can itself be converted to a carboxylic acid upon hydrolysis:

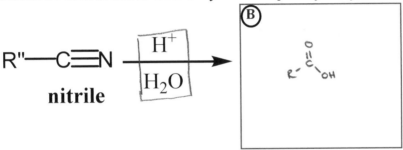

Notes

The mechanism of nitrile hydrolysis begins with nucleophilic addition to the polar CN bond and deprotonation to yield an imine. The imine hydrolysis is the revers of imine formation discussed earlier:

Notes

The α-position (α = alpha) is the site adjacent to the carbonyl carbon. This site can be deprotonated relatively easily because the conjugate base is stabilized by resonance:

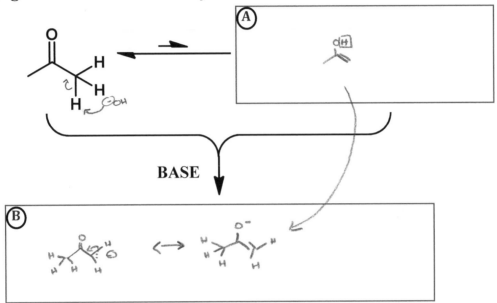

BASE

Notes

Lesson VI.15. Preparation of Enolates from Carbonyl Species
Stabilized enolates

When evaluating the relative stabilities of species, consider all influences. The enol form can be made more favorable, for example, if the alkene in the enol is part of a π-conjugated or aromatic system:

The α-position of an ester is not as acidic as the α-position of an aldehyde or ketone because its conjugate base is less stable. The lower stability of the ester enolate stems from a repulsive inductive effect:

acetone   NaOH

A

enolate

α-anion

steric repulsion

ethyl acetate   NaOCH₃

EtO   EtO

steric repulsion

need a stronger base to deprotonate an ester

Notes

187

Enolates can be used as nucleophiles, for example in an S_N2 reaction. This is sometimes called an α-alkylation reaction, since it adds an alkyl group to the carbonyl's α-position:

Mechanistically, the enolate is formed first:

redraw to
this form

★ the carbon should always have the
negative and be the nucleophile

Notes

There are two ways that you may see nucleophilic attack by an enolate represented, depending on which resonance contributor is represented:

Two ways to represent the same S_N2 reaction

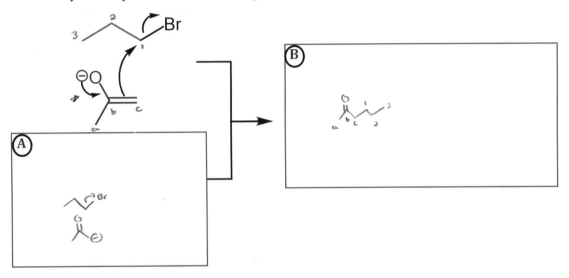

Notes

A deprotonated carbonyl can be used as a reagent for reactions other than Aldol condensation as well. One useful reaction is **α-halogenation:**

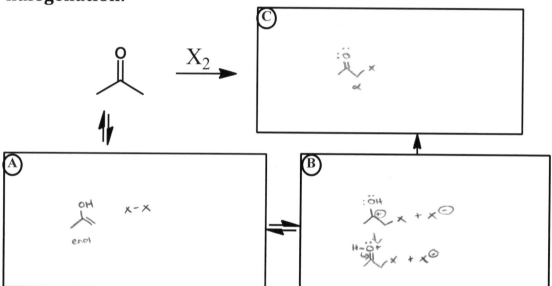

**Notes**

When more than one enolizable proton is present, more than one halogen can be added to the alpha carbon, eventually leading to what may initially be unanticipated products:

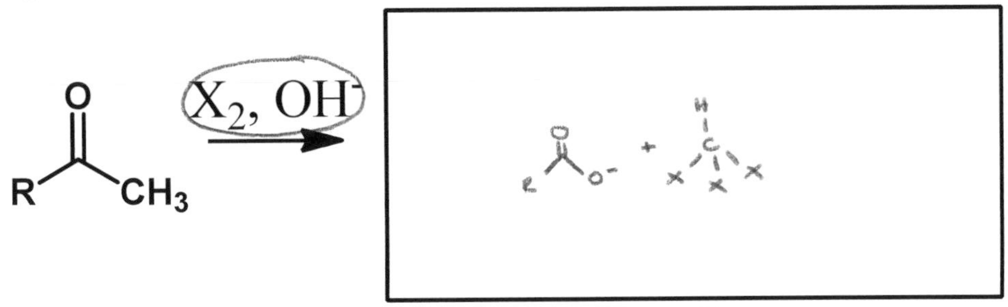

Balanced:

HCX_3 = haloform

$HCCl_3$ = chloroform

$HCBr_3$ = bromoform

Notes

If we examine the mechanism in light of our previous knowledge, the origin of the products should become clearer:

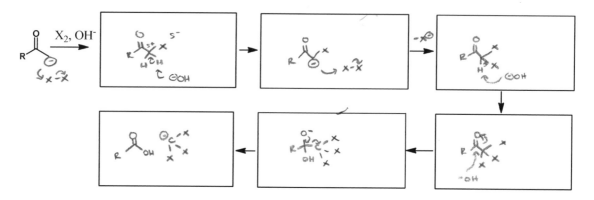

Notes

Lesson VI.17. Aldol Addition
Aldol addition is a Type A reaction

If an enolate is used as the nucleophile in a Type A reaction, this is called an **Aldol Addition**.

Notes

Intramolecular aldol addition can be used to form cyclic structures:

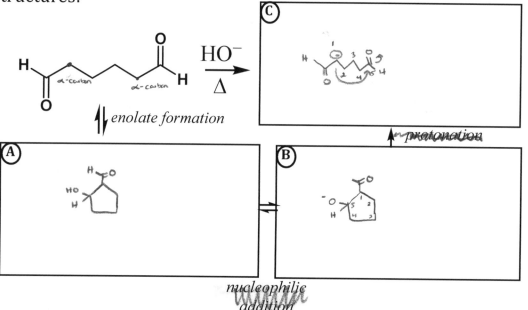

enolate formation

nucleophilic addition

Notes

Bicyclic structures can also be formed intramolecularly:

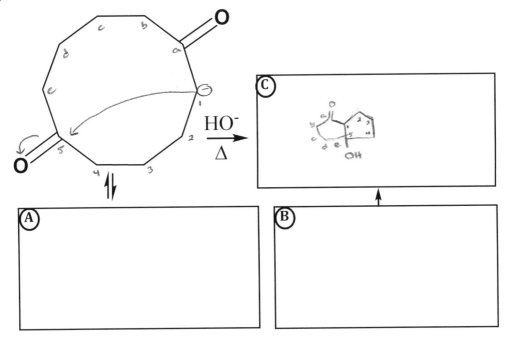

$$\xrightarrow[\Delta]{HO^-}$$

Ⓒ

Ⓐ

Ⓑ

<u>Notes</u>

Lesson VI.17. Aldol Addition
Mixed aldol reactions

Although the aldol addition is quite useful, it has its drawbacks. Consider, for example what would happen if you mixed two different carbonyls together to try to couple them:

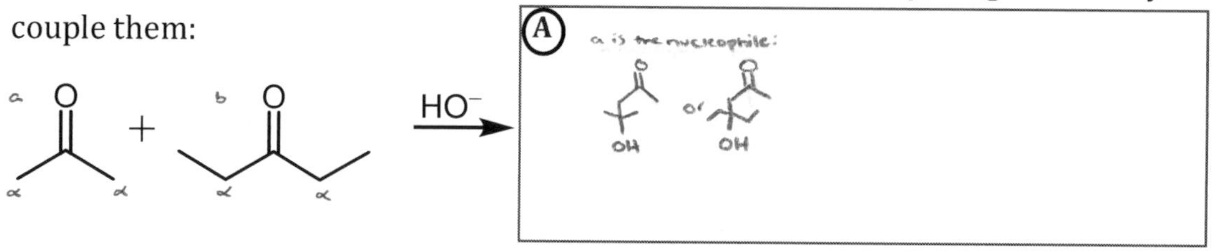

Or even one carbonyl that happens to have a different R group on each side of the carbonyl:

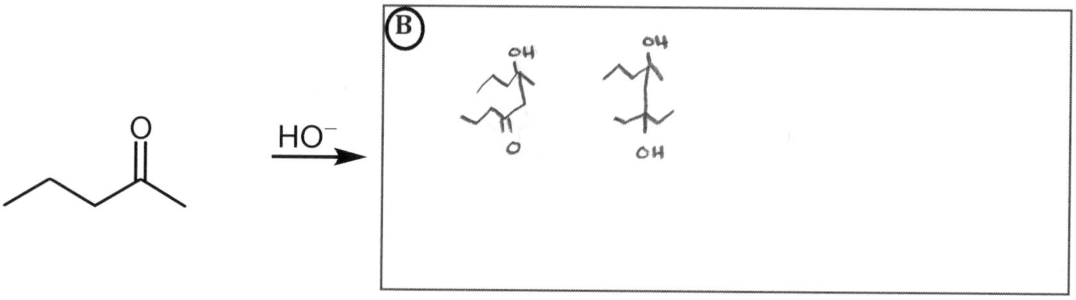

Notes

Because product mixtures result from the reactions shown on the previous page, the reaction is often cleanest when there is only **ONE type** of enolizable (alpha) proton available:

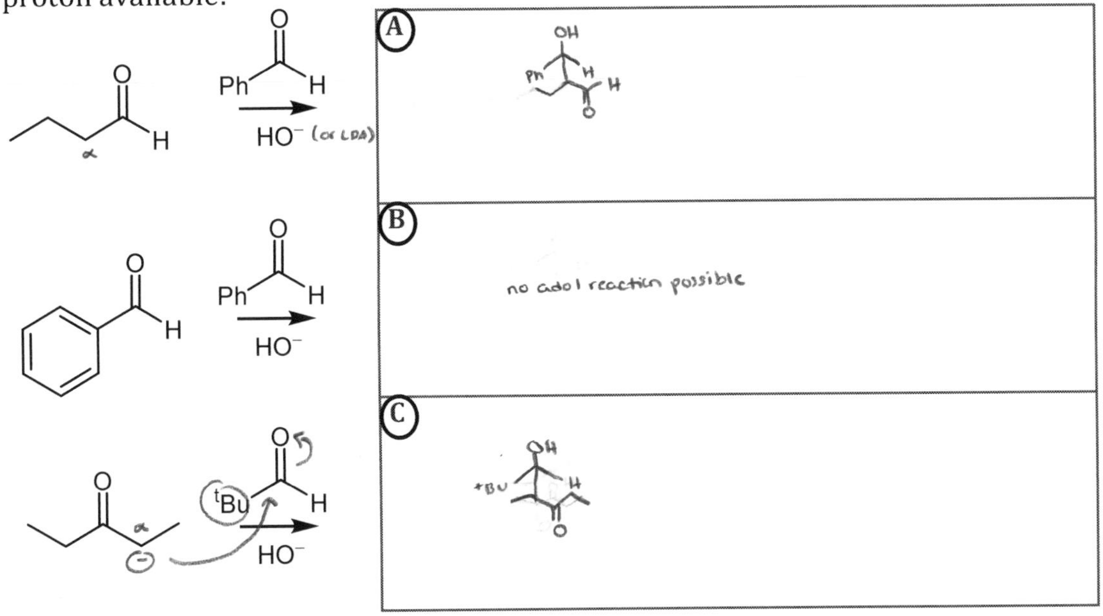

(A)

(B)   no adol reaction possible

(C)

Notes

We have seen that an enolate can serve as the nucleophile in a Type A reaction with aldehydes and ketones to give aldol addition products. With continued heating, the aldol addition product undergoes further reaction by an E2-like pathway to give the **aldol condensation product:**

Y = H : aldehyde
Y = R: ketone

Note that the alkene unit can exhibit *E-/Z-* and *cis-/trans-* isomers. The major alkene product formed is:

C
the most stable

Notes

example problems:

:N~R

When the nucleophile is an enolate generated by deprotonation of an ester, the particular S_NAc reaction is called the **Claisen Condensation**:

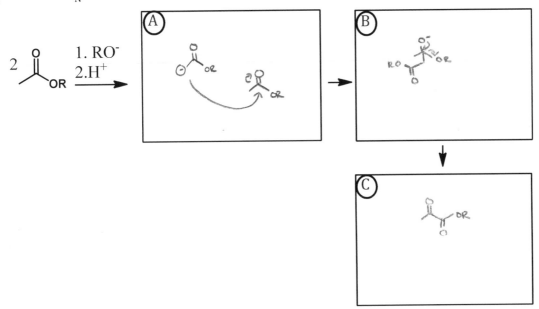

Lesson VI.18. Claisen Condensation
Intermolecular Claisen is called Dieckmann condensation

If a Claisen condensation happens to occur intramolecularly, this is called a
Dieckmann condensation:

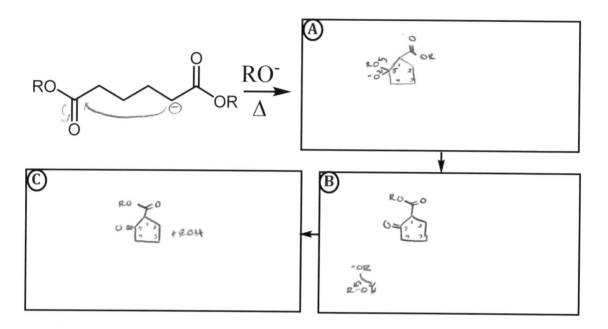

Notes

A reaction that is essentially the same as the Claisen reaction involves using an enolate from a ketone or aldehyde as the nucleophile to attack an ester's carbonyl carbon:

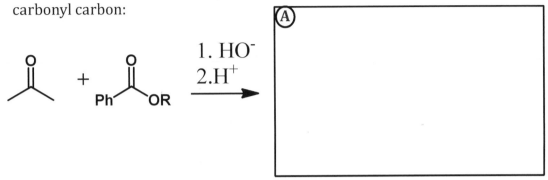

Note that the difference in **anion stability** of the enolate derived from a ketone or aldehyde versus that derived from an ester leads to a strong preference for deprotonation at the alpha carbon of the ketone or aldehyde:

Ⓑ

Notes

It is possible to remove not only the carbonyl O, but the entire C=O unit. This is done by a process called **decarboxylation**. This is often done simply by heating. The temperature required varies on the basis of the functional group. The 3-oxo-carboxylic acids are relatively easy to decarboxylate:

3-oxo carboxylic acid

tautomerization

carboxylic
acid

can lose CO_2

Now that we know that a 3-oxo carbonyl is often relatively easy to decarboxylate, we can exploit this property in some specific and very useful reactions. The first of these that we will cover is the **malonic ester synthesis**. This reaction typically starts with diethyl malonate and is used to prepare various carboxylic acids:

1. RO⁻

2. R'Br

3. HCl, H₂O, Δ

(A)

We should be able to reason through what's happening in these three steps.

Step 1. Base:

Step 2: R'Br:

Step 3. Acid and heat:

    a.

    b.

Notes

Lesson VI.19. Decarboxylation and Synthetic Applications
Malonic ester synthesis

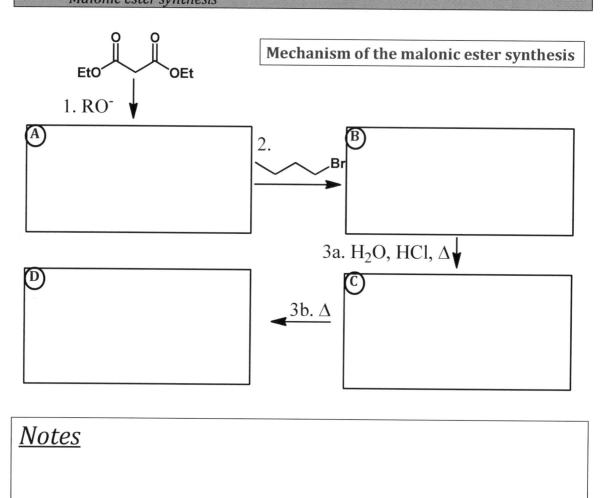

Mechanism of the malonic ester synthesis

1. RO⁻

A

2.

Br

B

3a. H₂O, HCl, Δ

C

3b. Δ

D

Notes

Now that we have seen the malonic ester synthesis, the **acetoacetic ester synthesis** should be relatively straightforward. This reaction simply has a methyl group in place of one of the two ethoxy groups of diethylmalonate:

The three steps are quite analogous to those we saw for malonic ester synthesis as well...

Step 1. Base:

Step 2: R'Br:

Step 3. Acid and heat:

     a.

     b.

Notes

We have seen enough examples to know how carbonyls react with nucleophiles and how an alkene might react with a nucleophile:

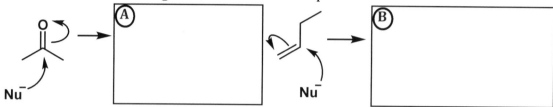

If the carbonyl and the alkene are in pi-conjugation with one another, what might the impact be?

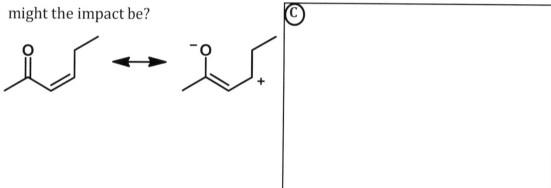

___

Notes

Lesson VI.20. Addition of Nucleophiles to α,β-Unsaturated Carbonyls

Direct addition versus conjugate addition in unsaturated carbonyls

As a result of the **two electrophilic sites** present in an α,β-unsaturated carbonyl, there are two potentially competing pathways for reaction with a nucleophile:

1,2-addition (direct addition)

(A)

1,4-addition (direct addition)

(B)

Notes

Fortunately, the type of addition depends in large extent on whether the attacking nucleophile is a strong or weak base. Strong bases do 1,2-addition.

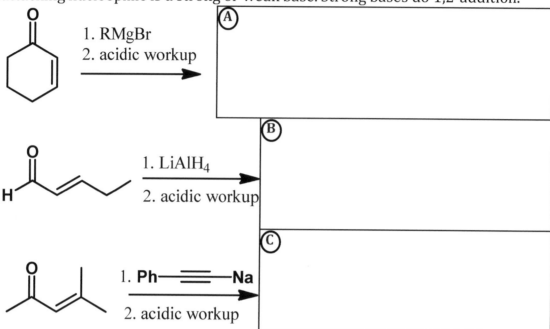

Notes

Weak bases (stable anions) tend to do 1,4-addition:

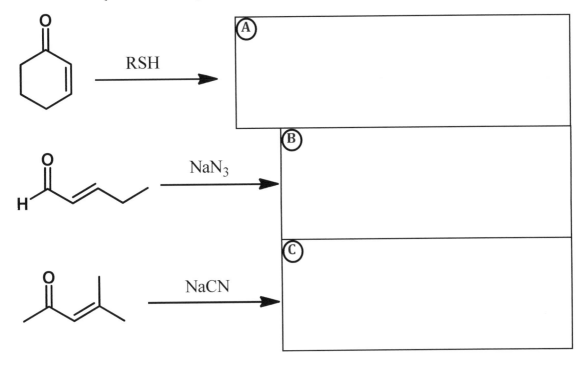

A

B

C

Notes

Let us use our newfound knowledge of how unsaturated carbonyls tend to react to predict what might happen if an unsaturated carbonyl reacts with an enolate. First, evaluate the enolate's relative base strength:

Ⓐ

Now we should be able to predict what an enolate nucleophile would tend to do upon reaction with an α,β–unsaturated carbonyl:

Ⓑ

Notes

Lesson VI.20. Addition of Nucleophiles to α,β-Unsaturated Carbonyls
Michael Addition

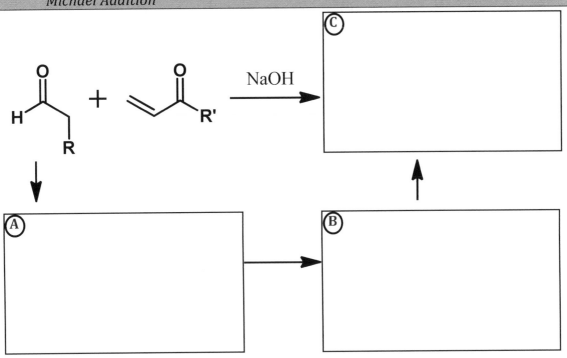

Notes

Lesson VI.20. Addition of Nucleophiles to α,β-Unsaturated Carbonyls
Michael Addition

The conditions for the Michael reaction are the same type of conditions used for the Aldol condensation. For this reason, the reactions can be in competition with each other. The Michael reaction can be made more favorable if we use β-diketones as the enolate component of a Michael addition:

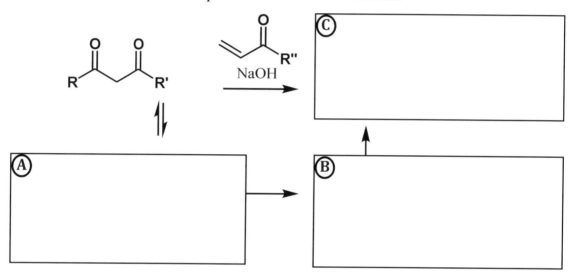

Notes

212

You may notice that the product of a Michael reaction is a species that contains at least two carbonyls in it. This means that we can do an intramolecular aldol condensation reaction *between* the carbonyls on the initial Michael product. This is called **Robinson Annulation**, an important route for making substituted cyclohexanones:

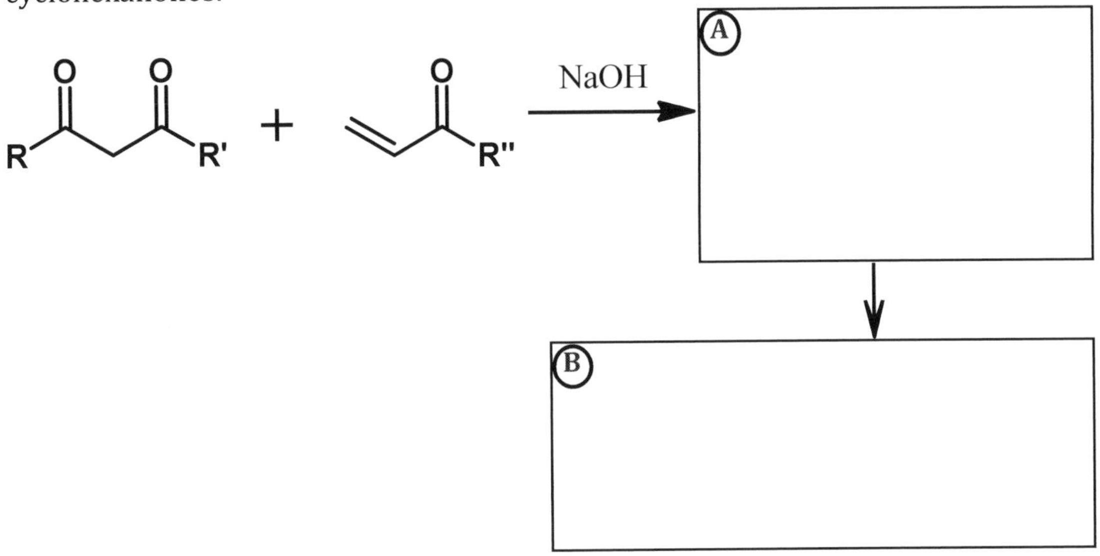

Notes

Lesson VI.20. Addition of Nucleophiles to α,β-Unsaturated Carbonyls
Robinson Annulation

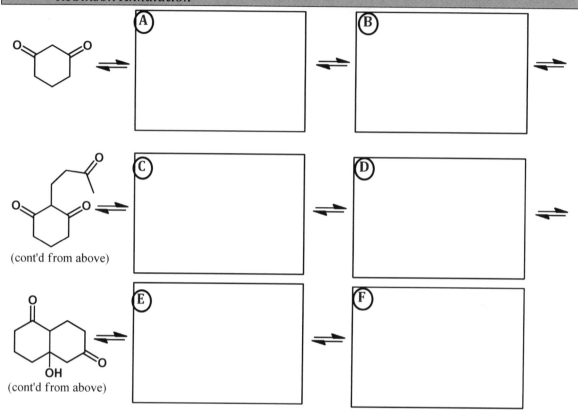

PART VII: Methods for Determining the Structure of Organic Compounds

Different energies of light elicit different changes in molecules that absorb them. In this course, we will consider UV, visible, IR and radio frequency radiation:

higher energy

⟵──────────────────────────────────────

lower wavelength

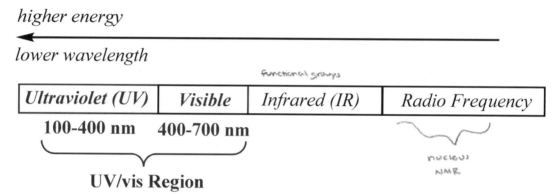

Ultraviolet (UV)	Visible	Infrared (IR)	Radio Frequency

100-400 nm   400-700 nm

UV/vis Region

First, we will focus on how UV and visible light, collectively abbreviated UV–visible (UV/vis) light, interact with organic molecules. The UV/vis part of the spectrum we will consider spans a wavelength range from ~100–700 nm.

Notes

When a molecule absorbs UV/vis radiation of an appropriate energy, it causes one of the electrons to undergo an **electronic transition** from the highest occupied molecular orbital (HOMO) to the lowest unoccupied molecular orbital (LUMO).

An electron in a σ-bond gets promoted to a **σ-antibonding (σ*) orbital**:

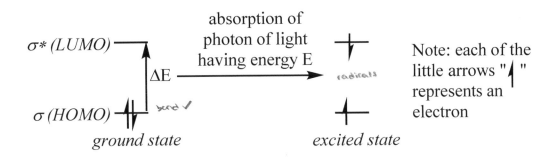

σ* (LUMO) ————

ΔE

σ (HOMO) —— bond ✓

ground state

absorption of
photon of light
having energy E

ΔE ————————→ radicals

excited state

Note: each of the
little arrows "↿"
represents an
electron

Notes

- C̈–C̈
  C̈   C̈

An electron in a π-bond gets promoted to a **π-antibonding (π*) orbital**:

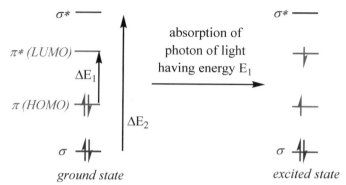

ground state   excited state

Compared to the σ→ σ* transition, the π→ π* is:

(A)

lower energy (to break a π bond)

Furthermore, the longer the π conjugated system, the lower the energy of the photon needed to promote the π→π* transition.

(B)

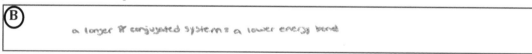

a longer π conjugated system = a lower energy bond

Notes

218

Lesson VII.2. UV-vis Spectroscopy
UV/vis spectrometer

A UV-vis spectrometer is set up as follows:

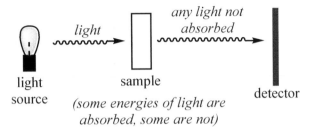

The spectrum is a plot of absorbance versus wavelength:

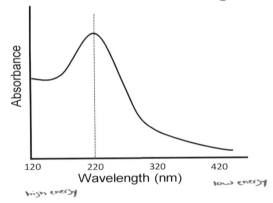

Notes

219

The amount of light absorbed per mole of a sample is called the **molar absorptivity** or **molar extinction coefficient**.

The **Beer-Lambert Law** provides an equation relating the absorbance (A), pathlength (b), concentration (c) and extinction coefficient (ε):

For a constant pathlength, then, it is easy to monitor concentration, and thus to follow the reaction rate. If a species we are following gives spectrum **A** (absorbing at 220 nm) at the start of a reaction, and spectrum **B** after 1 h, we know that half of the compound is consumed in that one hour, because the absorption is halved.

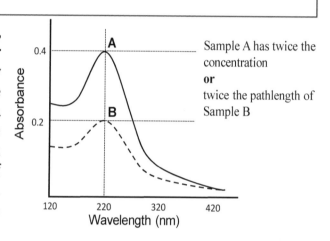

Sample A has twice the concentration
or
twice the pathlength of Sample B

Notes

Molecules absorb energy:

infrared light

Leading to:

bond stretching and bond angle bending

Expressing energy: $E = h\upsilon = \dfrac{hc}{\lambda}$

wavenumber ($\tilde{\upsilon}$):

cm⁻¹ (reciprocal centimeters)

Larger wavenumbers =

higher energy

Notes

Lesson VII.3. Interaction of Infrared Light with Molecules
Vibrational modes; stretching modes and bending modes

Bonds can vibrate in different ways, and each of these **vibrational modes** requires a different energy:

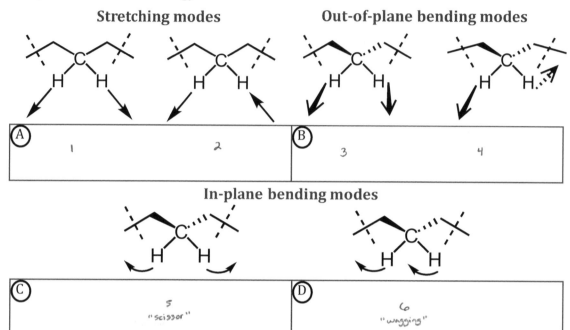

Stretching modes

Ⓐ 1   2

Out-of-plane bending modes

Ⓑ 3   4

In-plane bending modes

Ⓒ 5
"scissor"

Ⓓ 6
"wagging"

Notes

A typical IR spectrum has this appearance:

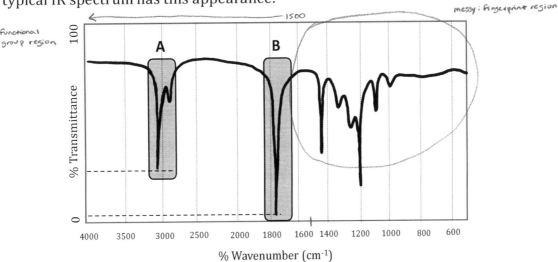

% Wavenumber (cm⁻¹)

Band **A** corresponds to a set of bonds that absorbs ~75% of the IR light at 3000 cm⁻¹ emitted by the source (i.e., 25% transmittance), whereas band **B** corresponds to a set of bonds that absorbs nearly 95% of the IR light at 1750 cm⁻¹ (i.e., 5% transmittance).

Notes

223

Each type of bond has a characteristic energy of absorption for IR radiation:

Bond	Energy (cm^{-1})	Intensity
N≡C	2255-2220	m-s
C≡C	2260-2100	w-m
C=C	1675-1660	m
N=C	1650-1550	m
⬡ {	1600 **AND**	w-s
	1500-1425	
C=O	1775-1650	s
C—O	1250-1000	s
C—N	1230-1000	m
O—H	3650-3200	s (br)
O—H	3300-2500	s (br)
N—H	3500-3300	m (br)
C—H	3300-2725	m

C-H Bond (Stretch) — Energy (cm^{-1})

	Energy (cm^{-1})
C≡C—H	3300-ish
C=C—H	3100-3000
C—C—H	2950-2850

O=C(CH₃)(H) : 2820-ish and 2720-ish

C-H Bond (Bending)

—CH₃
—CH₂—   —C(H)— } 1450-1400

980-960 trans

730-670 cis

840-800 trisubstituted

990 and 910 monosubstituted

890 disubstituted terminal

Notes

There are several points worth noting:

Stronger bonds:

 (A)

take more energy to stretch

More polar bonds:

 (B)

- Stronger than comparable less polar bonds
  - comparable: single versus single, double versus double, etc.

Resonance influences bond order, and therefore bonds strength:

Resonance Contributors

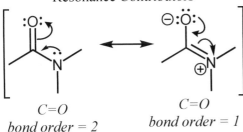

$C=O$
bond order = 2

$C=O$
bond order = 1

amide

Resonance Hybrid

 (C)

Amides have lower-energy C=O stretches than ketones

Notes

Lesson VII.4. Infrared Spectroscopy
IR decision tree

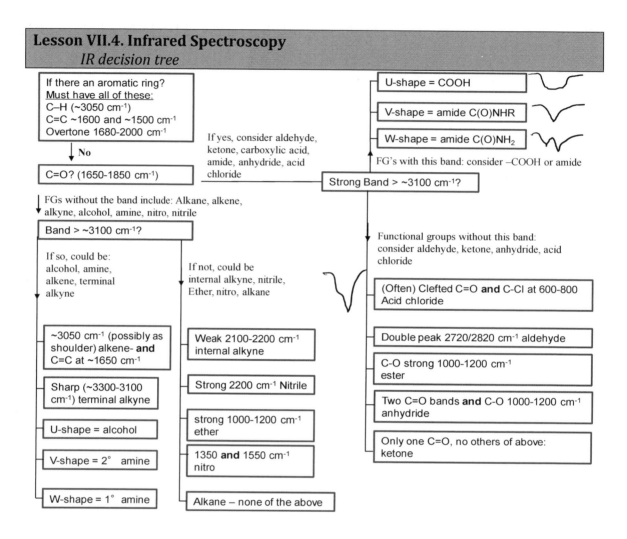

If there an aromatic ring?
<u>Must have all of these:</u>
C–H (~3050 cm^{-1})
C=C ~1600 and ~1500 cm^{-1}
Overtone 1680-2000 cm^{-1}

↓ **No**

C=O? (1650-1850 cm^{-1})

If yes, consider aldehyde, ketone, carboxylic acid, amide, anhydride, acid chloride

U-shape = COOH

V-shape = amide C(O)NHR

W-shape = amide C(O)NH$_2$

FG's with this band: consider –COOH or amide

Strong Band > ~3100 cm^{-1}?

↓ FGs without the band include: Alkane, alkene, alkyne, alcohol, amine, nitro, nitrile

Band > ~3100 cm^{-1}?

If so, could be: alcohol, amine, alkene, terminal alkyne

If not, could be internal alkyne, nitrile, Ether, nitro, alkane

Functional groups without this band: consider aldehyde, ketone, anhydride, acid chloride

~3050 cm^{-1} (possibly as shoulder) alkene- **and** C=C at ~1650 cm^{-1}

Weak 2100-2200 cm^{-1} internal alkyne

(Often) Clefted C=O **and** C-Cl at 600-800 Acid chloride

Sharp (~3300-3100 cm^{-1}) terminal alkyne

Strong 2200 cm^{-1} Nitrile

Double peak 2720/2820 cm^{-1} aldehyde

U-shape = alcohol

strong 1000-1200 cm^{-1} ether

C-O strong 1000-1200 cm^{-1} ester

V-shape = 2° amine

1350 **and** 1550 cm^{-1} nitro

Two C=O bands **and** C-O 1000-1200 cm^{-1} anhydride

W-shape = 1° amine

Alkane – none of the above

Only one C=O, no others of above: ketone

Notes

IR Spectrum for Hexanal

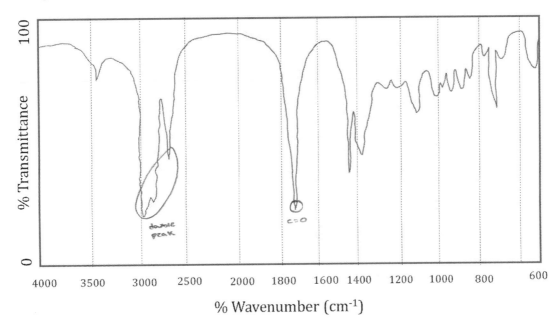

Notes

· hexanal

· only the aldehyde shows a peak between 2500-2700ish

IR Spectrum for cyclohexene

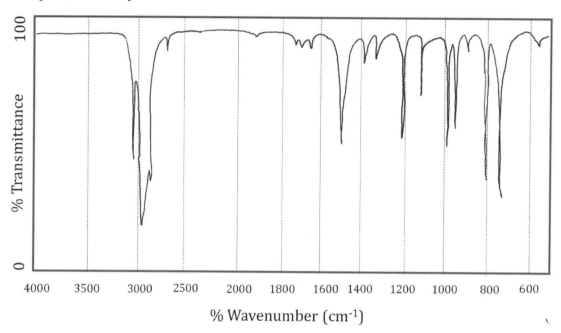

% Wavenumber (cm^{-1})

Notes

IR Spectrum for nitrohexane

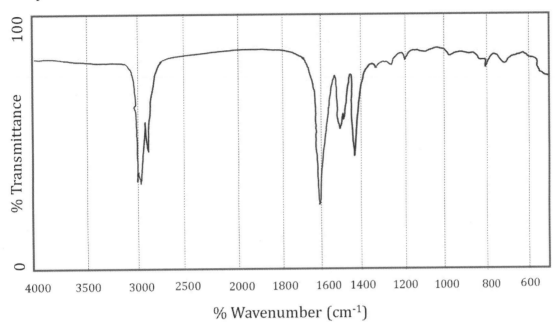

% Wavenumber (cm^{-1})

<u>*Notes*</u>

IR Spectrum for 1-hexyne

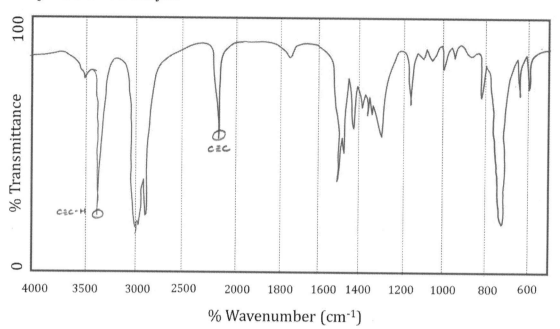

% Wavenumber (cm⁻¹)

Notes

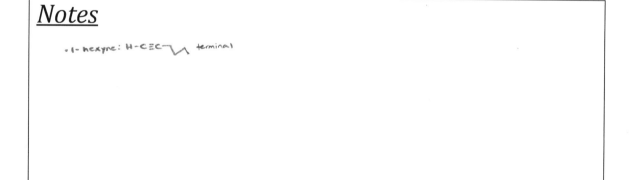

IR Spectrum for 2-butyne

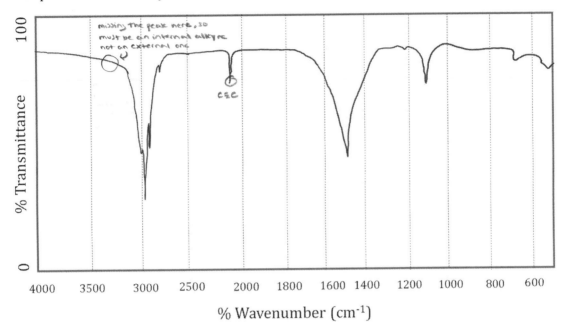

Notes

. 2-butyl: —≡— internal alkyne

Lesson VII.4. Infrared Spectroscopy
IR Practice

IR Spectrum for 3-octanol

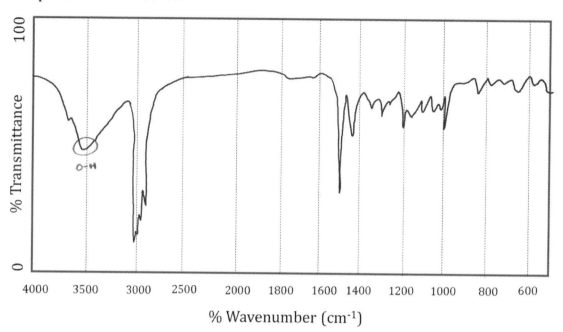

Notes

Lesson VII.4. Infrared Spectroscopy
IR Practice

IR Spectrum for diethylamine

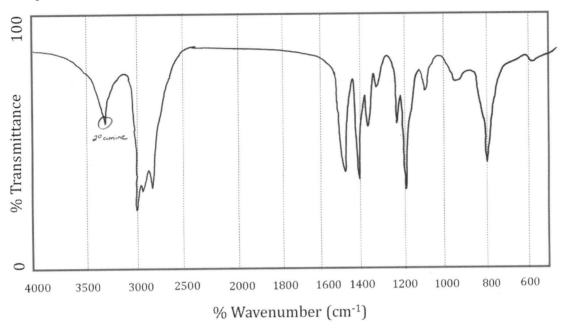

Notes

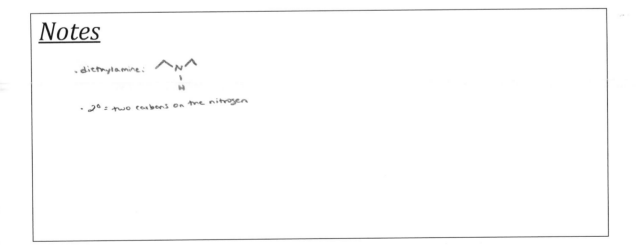

- diethylamine: (structure)
- 2° = two carbons on the nitrogen

Lesson VII.4. Infrared Spectroscopy
IR Practice

IR Spectrum for hexylamine

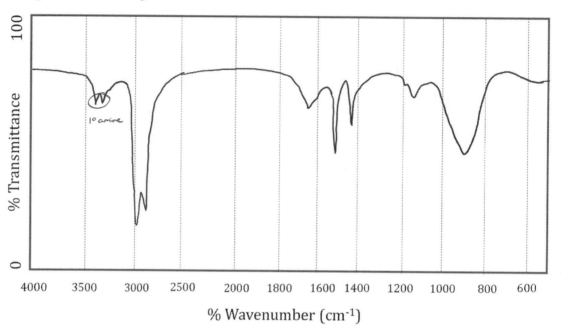

% Wavenumber (cm⁻¹)

Notes

- hexylamine:
- 1°: one carbon on the nitrogen

234

IR Spectrum for acetonitrile (CH_3CN)

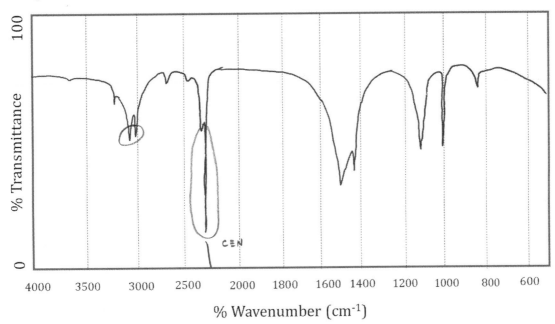

C≡N

% Wavenumber (cm^{-1})

Notes

- acetonitrile:
- the "w" is not greater than 3100, so it is not a 1° amine

IR Spectrum for diethylether

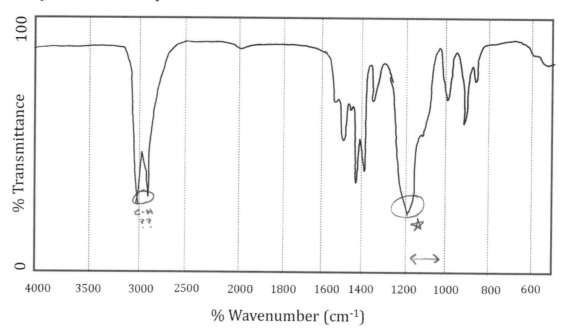

Notes

IR Spectrum for hexanoic acid

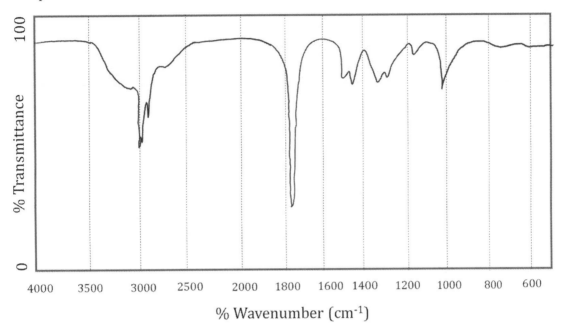

% Wavenumber (cm⁻¹)

Notes

IR Spectrum for *N*-methylacetamide $(CH_3C(O)N(H)CH_3)$

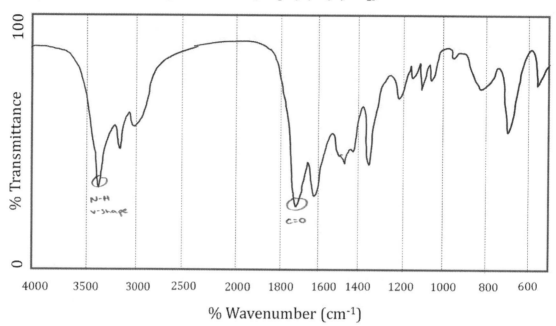

% Wavenumber (cm^{-1})

Notes

- *N*-methylacetamide :

IR Spectrum for propamide $(CH_3CH_2C(O)NH_2)$

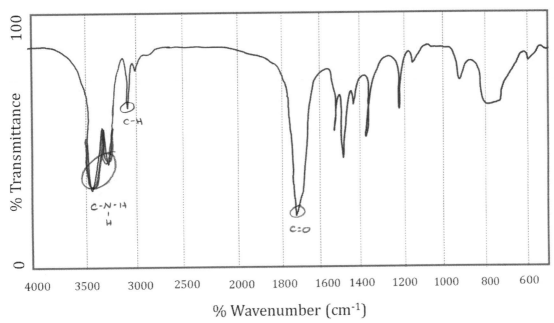

Notes

• propamide: (structure drawn)

skip

Lesson VII.4. Infrared Spectroscopy
IR Practice

IR Spectrum for propanoyl chloride

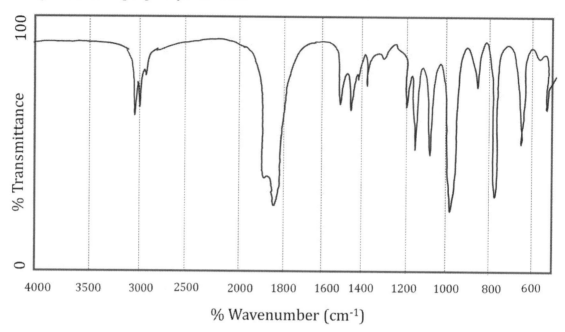

% Wavenumber (cm^{-1})

Notes

IR Spectrum for acetic anhydride

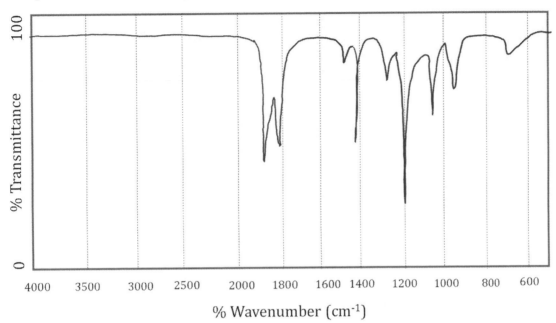

% Wavenumber (cm⁻¹)

Notes

 acetic anhydride:

Lesson VII.4. Infrared Spectroscopy
IR Practice

IR Spectrum for ethylbenzene

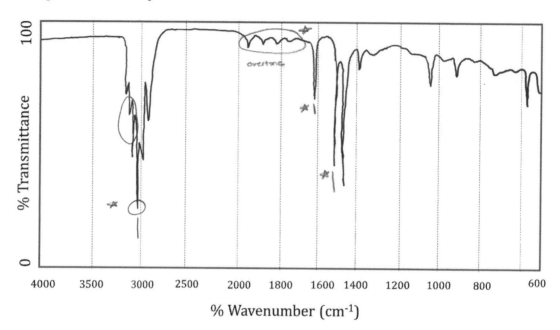

Notes

ethylbenzene :

NMR stands for:

Ⓐ

nuclear magnetic resonance

NMR spectroscopy is a technique that is used to identify compounds.
A NMR Spectrum is a plot of

Ⓑ

energy (x-axis)

vs.

Ⓒ

intensity (y-axis)

Many common nuclei are NMR active, including:

Ⓓ

^1H, ^{13}C

In this Lecture Guide, we will focus on ^{13}C and ^1H NMR spectrometry

Notes

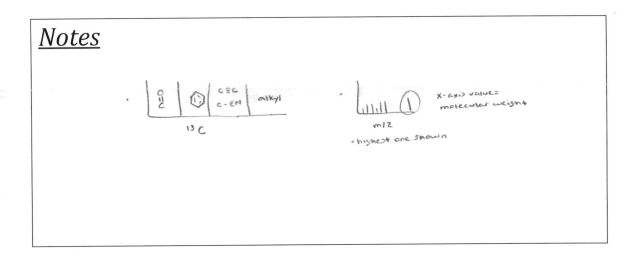

A representative spectrum is shown here:

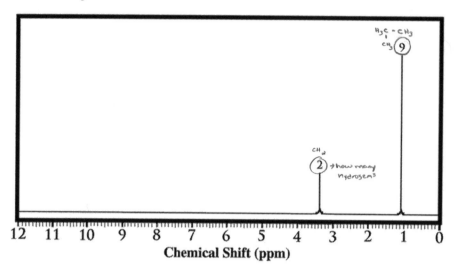

Chemical Shift (ppm)

The energy at which we observe a peak can tell us

(D) what type of carbon or hydrogen we have

thus aiding in the compound's identification.

Notes

Nuclei are charged. Charged particles interact with magnetic fields. A 'resonating' nucleus generates a magnetic field of its own. This generated magnetic field may be aligned with or oppose the applied field. It takes more energy to oppose the applied field.

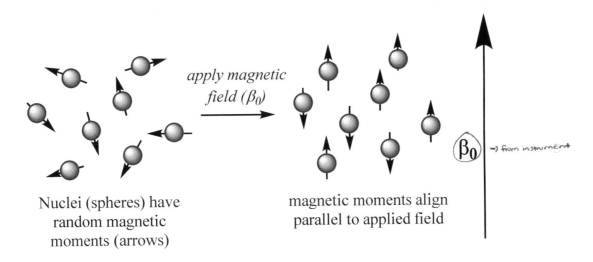

apply magnetic field (β_0)

β_0 → from instrument

Nuclei (spheres) have random magnetic moments (arrows)

magnetic moments align parallel to applied field

Notes

One can add energy to get the nucleus' magnetic field to 'flip' direction. By measuring the energy needed to accomplish this flip, we generate a NMR spectrum:

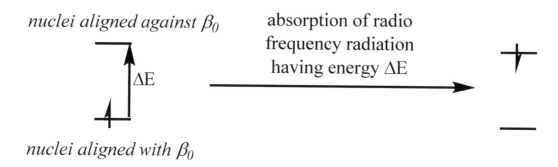

nuclei aligned against β_0

absorption of radio frequency radiation having energy ΔE

ΔE

nuclei aligned with β_0

Notes

Each set of **magnetically equivalent** nuclei absorb a certain energy of photon. Free rotation about single bonds averages the signal, so that nuclei that can be interconverted by single bond rotation are equivalent. Consider how many types of H nuclei are in each of these molecules:

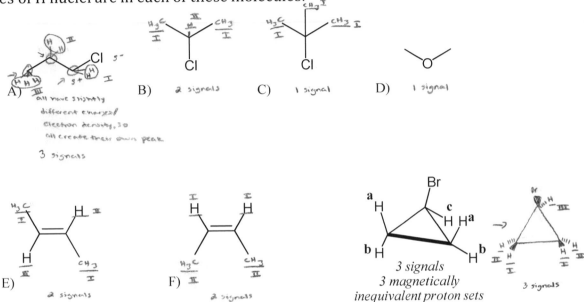

A) all have slightly different energies/ electron density, so all create their own peak.

3 signals

B) 2 signals

C) 1 signal

D) 1 signal

E) 2 signals

F) 2 signals

3 signals
3 magnetically inequivalent proton sets

3 signals

Lesson VII.5. Introduction to Nuclear Magnetic Resonance
Shielding by electrons in NMR spectra

Nuclei in molecules have electrons around them. Since electrons are oppositely charged compared to nuclei, they exert an opposing effect on the applied magnetic field. More electron density thus **shields** a nucleus and lowers the energy needed to flip it. In this way, NMR is an indirect way to measure electron density, which allows us to deduce the type of group containing the nucleus.

Protons in Electron Poor environment: **Protons in Electron Rich environment:**

Frequency increases

δ **Increases**

Higher number on x axis in spectrum

Notes

The area under each signal in an NMR spectrum is proportional to the number of nuclei giving the signal. The area under the peak is called the **integration.**

The integrations are printed above the peaks in this book. Consider this ^{1}H NMR spectrum:

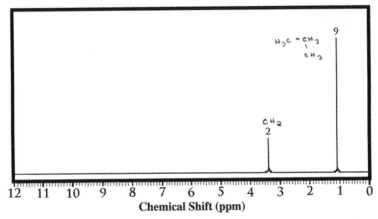

The ratio of peaks is 2:9. This could mean that one peak is attributable to 2 H nuclei and the other is attributable to 9, or one to 4 H and one to 9 H; we only know the ratio.

Notes

$C_5H_{11}Cl$ give structure:

$$Cl - \overset{\overset{\displaystyle H_3C}{|}}{\underset{\underset{\displaystyle H}{|}}{C}} - \overset{\overset{\displaystyle}{|}}{\underset{\underset{\displaystyle CH_3}{|}}{C}} - CH_3$$

Lesson VII.5. Introduction to Nuclear Magnetic Resonance
Splitting and multiplicity in NMR

If there is an NMR active nucleus (i.e., one which resonates thus creating a magnetic field) near another NMR-active nucleus, the two will influence each other. Depending on the direction of nucleus **A**'s magnetic field, nucleus **A** may shield or reinforce the magnetic field experienced by its neighboring nucleus **B**:

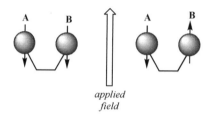

applied field

This causes the signal for nucleus **A** to split slightly into two peaks. The shape of the peak will be:

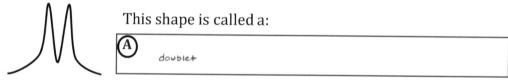

This shape is called a:

Ⓐ   doublet

As there are more and more NMR-active nuclei adjacent to nucleus **A**, the splitting gets more elaborate ...

<u>Notes</u>

For 1H NMR, the **multiplicity** (m, the number of smaller peaks into which the signal is split) is equal to $n+1$, where n is the number of H on neighbor C:

#H on neighbor C:	Type of peak		Ratio of heights	
0	singlet		1	$m=n+1$ $m=0+1$
1	doublet		1:1	
2	triplet		1:2:1	
3	quartet		1:3:3:1	
4	quintet		1:4:6:4:1	
5	sextet		1:5:10:10:5:1	
6	septet		1:6:15:20:15:6:1	

Notes

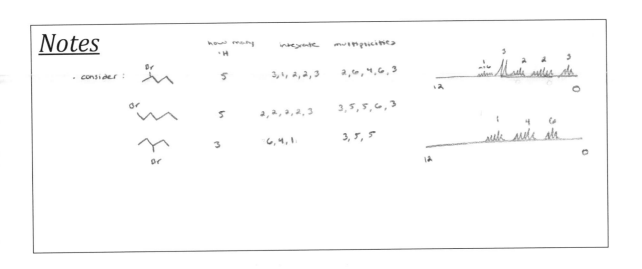

· consider :

	how many ·H	integrate	multiplicities
	5	3,1,2,2,3	2,6,4,6,3
or	5	2,2,2,2,3	3,5,5,6,3
or	3	6,4,1,	3,5,5
or			

General trends in ^{13}C NMR shifts are provided here:

Carbon (Shown)	Chemical Shift
$Si(CH_3)_4$	0
$—CH_3$	10-35
$—CH_2—$	15-50
$-\overset{H}{\underset{\vert}{C}}-$	20-60
$-\overset{\vert}{\underset{\vert}{C}}-$	30-40
$=C\overset{/}{\underset{\backslash}{}}$	100-150
benzene—H	110-175
$\equiv C—$	60-85

Carbon (Shown)	Chemical Shift	
$O—C$	50-80	
$N—C$	40-60	
$C—X$	$X = I$	0-35
	$X = Br$	20-65
	$X = Cl$	35-80
acetyl—Y	$\underline{Y =}$	
	H	190-200
	R	200-220
	OR	160-180
	OH	175-185
	NR_2	165-175

Notes

Carbon-13 NMR typical shifts in a more user-friendly format:

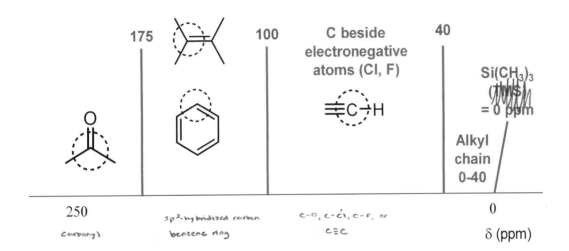

Notes

Carbon-13 NMR is usually run in such a way as to avoid any splitting, so the signals tend to all be singlets. Using our knowledge of the usual ranges for ^{13}C NMR resonances, we can infer some structural information from this spectrum:

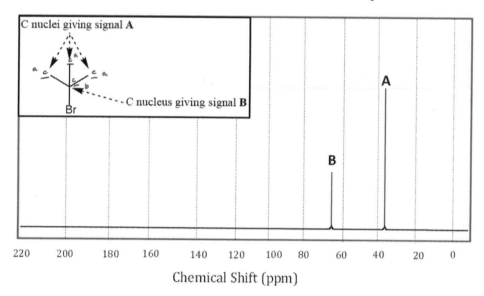

Chemical Shift (ppm)

Lesson VII.7. Proton Nuclear Magnetic Resonance Spectrometry
Interpreting proton NMR spectra

Let us examine what we can learn from each part of a ^1H NMR spectrum:

Chemical Shift tells us:

(A) how close to an electronegative atom or hybridization

Integration tells us:

(B) how many hydrogen (none in ^{13}C)

Caution:

Multiplicity tells us:

(C) how many hydrogen on neighboring carbons

Notes

General trends in ^{1}H NMR shifts are provided here:

Protons (Shown)	Chemical Shift
Si(CH$_3$)$_4$	0
—CH$_3$	0.9
—CH$_2$—	1.2
—C— (H)	1.4
CH$_3$ (allyl)	1.7
C(=O)CH$_3$	2.1
phenyl—CH$_3$	2.3
≡—H	2.4

Protons (Shown)	Chemical Shift
R—OCH$_3$	3.3
vinyl H	4.5-5.5
H—C—X, X = I	2.5-4
X = Br	2.5-4
X = Cl	3-4
X = F	4-4.5
phenyl—H	6.5-8.0
C(=O)—H	10

Notes

Proton NMR typical shifts in a more user-friendly format:

Integration:
H giving signal
Multiplicity:
Minus one to get
H on C's beside
site

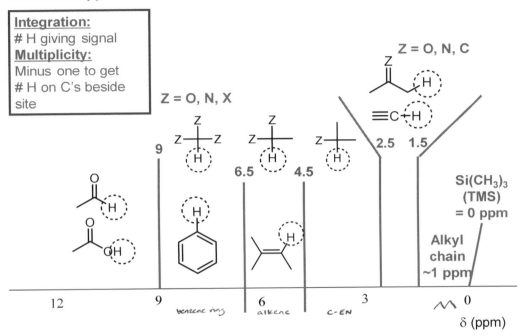

Notes

Appendix: Review of Select Topics from Organic Chemistry 1

Appendix: Review of Select Topics from Organic Chemistry 1
Acid-base strength and anion stability

$$\underset{\textit{acid}}{HA} + H_2O \rightleftharpoons \underset{\textit{conjugate base}}{A^-} + H_3O^+$$

Any factors that favor dissociation of HA into H^+ (to form H_3O^+ in water) and A^- will enhance acidity.

Most important factor influencing dissociation:

***1.** ______________________________________

Notes

The electronegativity plays a strong role because:

CH_4 $\qquad$ NH_3 $\qquad$ H_2O $\qquad$ HF

pK_a ~ _____ _____ _____ _____

Notes

If the atom to be deprotonated is the same and only hybridization changes, we need to know that the hybridization influences electronegativity:

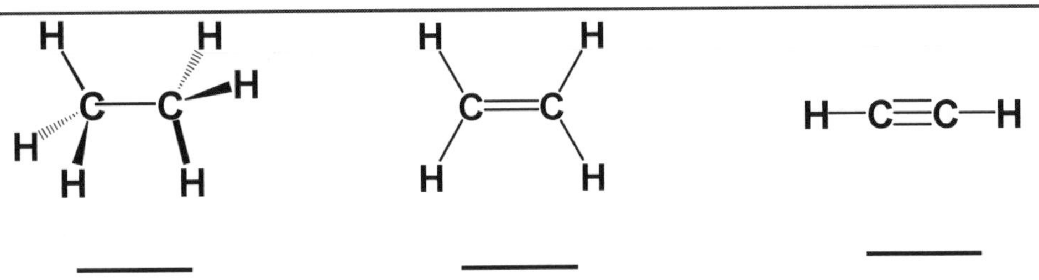

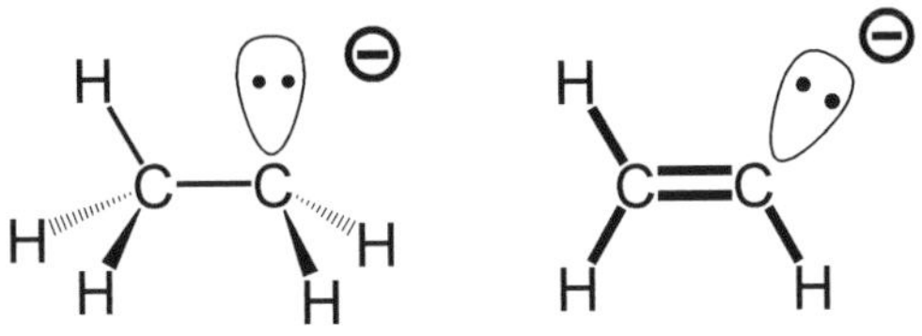

Note: the electronegativity of an *sp*-hybridized C is similar to that of an sp^3-hybridized N

Acetylide anion

Notes

Appendix: Review of Select Topics from Organic Chemistry 1

Acid-base strength and anion stability

If the anion produced by deprotonation has more than one (good) resonance form, then:

<u>HF</u>
pK_a = 3.1

<u>CH_3CO_2H</u>
pK_a = 4.7

<u>CH_3OH</u>
pK_a = 15.5

<u>Notes</u>

Within a group (column), the size of the anion has a strong effect on acidity, because:

_____ **HF** $pK_a \sim$ _____

HCl   _____

HBr   _____

_____ **HI**   _____

- ORBITAL OVERLAP
- **ANION STABILITY**

Notes

Inductive effects can make a species more OR less acidic. If an atom to be deprotonated has a partial positive charge INDUCED on it by nearby atoms, it is easier to deprotonate because:

pKa   4.8   3.2   2.9   2.8   2.7

This series illustrates:

pKa   Br 3.0   4.0   4.6

This series illustrates:

Notes

sterics:

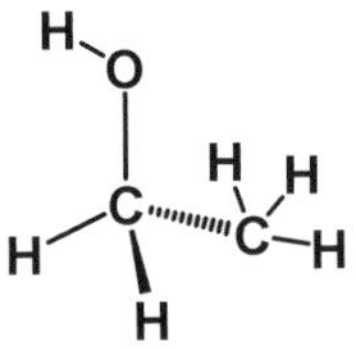

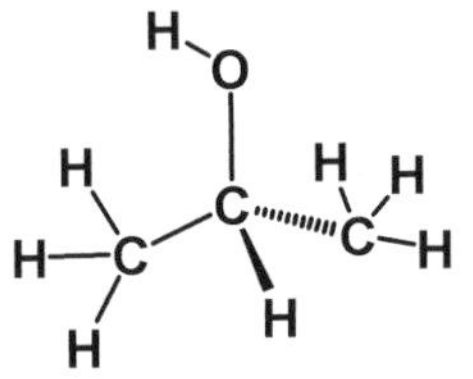

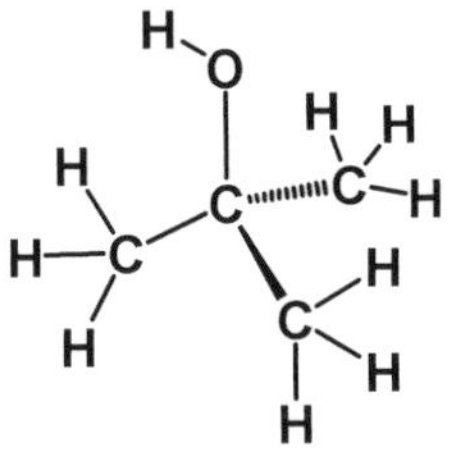

$pK_a \sim$ _____   _____   _____

_____

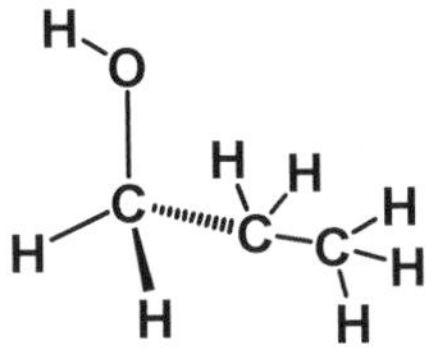

<u>Notes</u>

(A) Observation

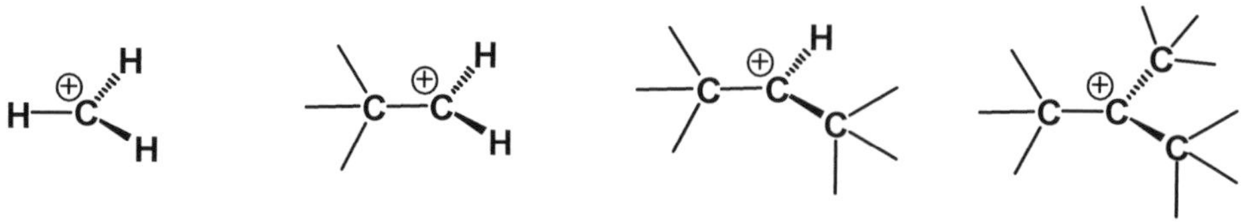

methyl 1^0 2^0 3^0

STABILITY

LESS STABLE   **MORE STABLE**

(B) Explanation: Hyperconjugation

<u>*Notes*</u>

Appendix: Review of Select Topics from Organic Chemistry 1
Cation stability

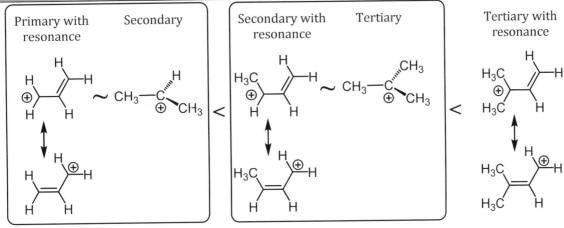

LESS STABLE

STABILITY

MORE STABLE

<u>Notes</u>

Appendix: Review of Select Topics from Organic Chemistry 1
Radical stability

Ⓐ **Observation**

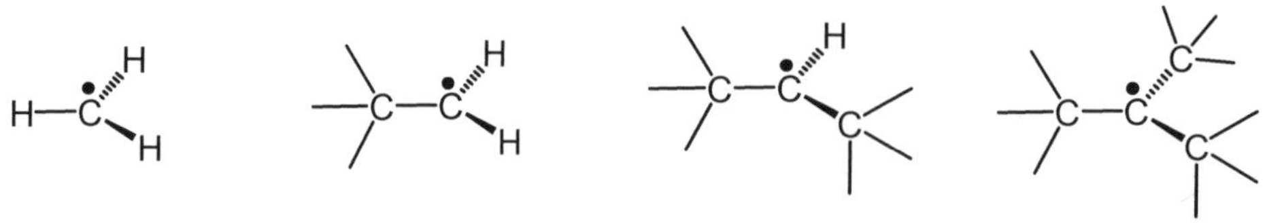

methyl   1°   2°   3°

LESS STABLE   **STABILITY**   **MORE STABLE**

Ⓑ **Explanation**

<u>Notes</u>

Appendix: Review of Select Topics from Organic Chemistry 1
Radical stability

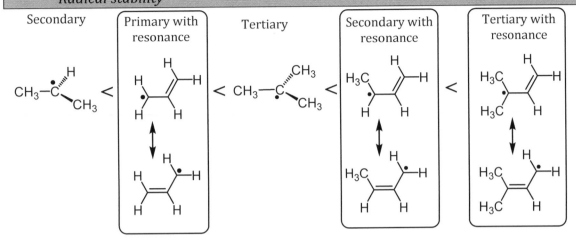

Secondary   Primary with resonance   Tertiary   Secondary with resonance   Tertiary with resonance

STABILITY

LESS STABLE   **MORE STABLE**

Notes

Thermodynamically Favorable (Spontaneous):

Example: Assuming that each process is mechanistically accessible, use your knowledge of stability trends to predict whether formation of product would be thermodynamically favorable (spontaneous):

A)

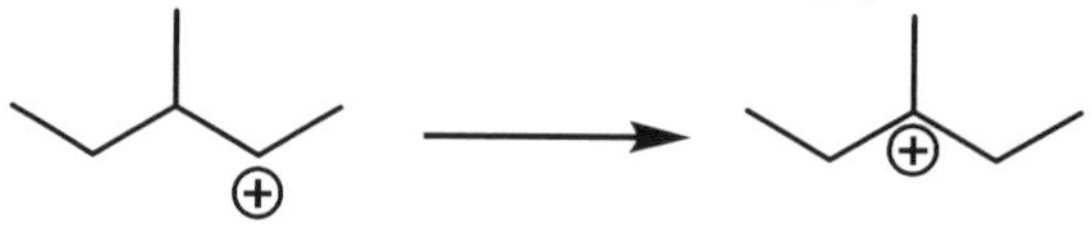

B) OH^- +

$\longrightarrow$ H_2O +

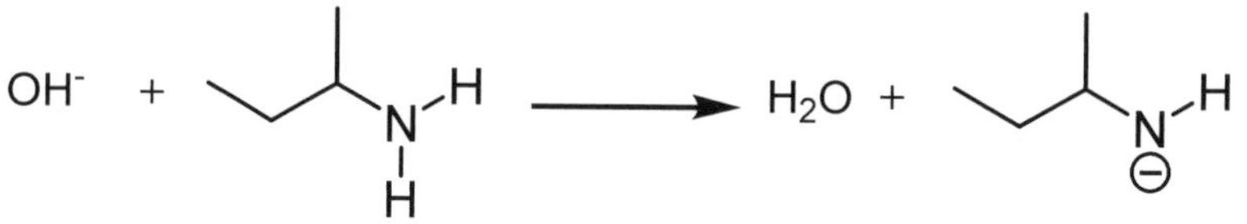

C) H^- +

$\longrightarrow$ H_2 +

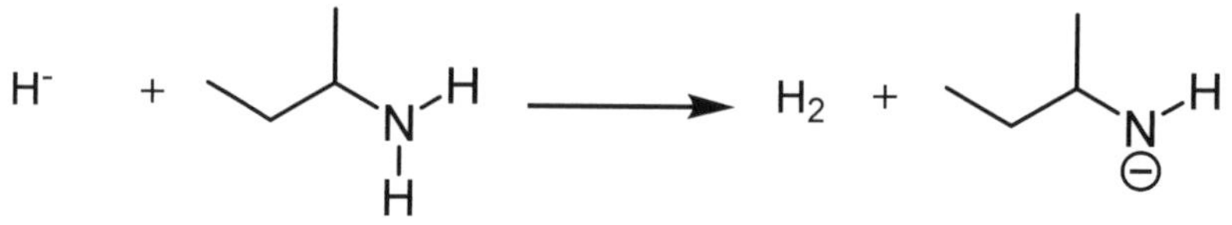

Notes

Definition of Equilibrium:

Equations:

Nucleophilic addition example :

S_N2 example:

Nucleophilic Elimination example:

Notes

Appendix: Review of Select Topics from Organic Chemistry 1

Reaction coordinate diagrams

Stability

Spontaneity:

Reaction Rate:

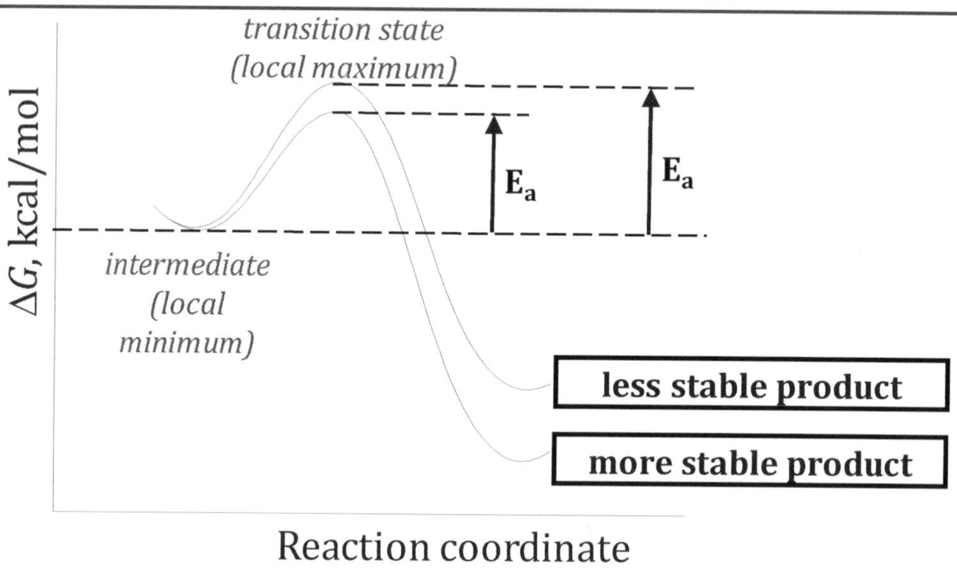
transition state
(local maximum)
ΔG, kcal/mol
E_a
E_a
intermediate
(local
minimum)
less stable product
more stable product
Reaction coordinate

Notes

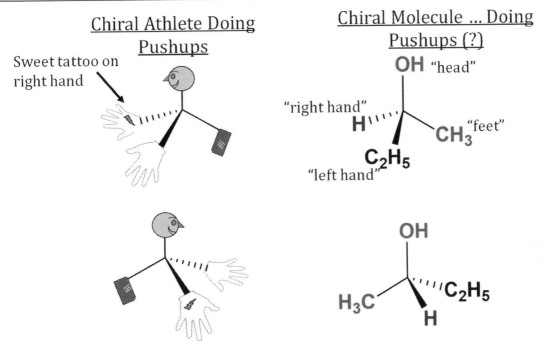

Chiral Athlete Doing Pushups

Sweet tattoo on right hand

Chiral Molecule ... Doing Pushups (?)

OH "head"

"right hand" H

"left hand" C₂H₅

CH₃ "feet"

OH

H₃C C₂H₅

H

We need a way to visualize rotation of a chiral molecule without losing stereochemical information

---

<u>Notes</u>

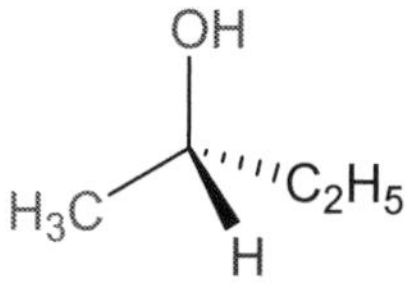

OH
H₃C ⟋⟍ ""C₂H₅
H

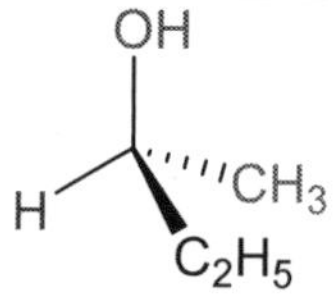

OH
H ⟋⟍ ""CH₃
C₂H₅

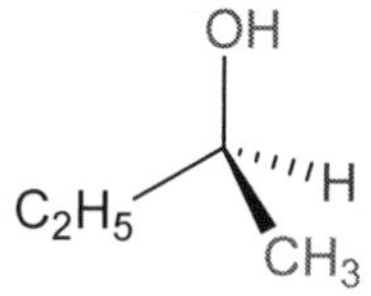

OH
C₂H₅ ⟋⟍ ""H
CH₃

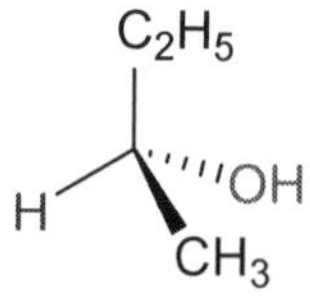

C₂H₅
H ⟋⟍ ""OH
CH₃

(A) Two other ways:

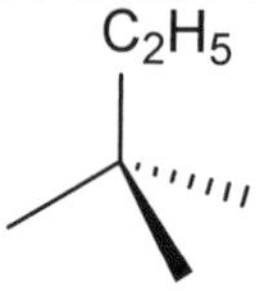

C₂H₅

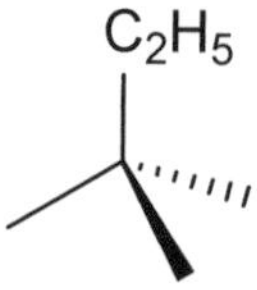

C₂H₅

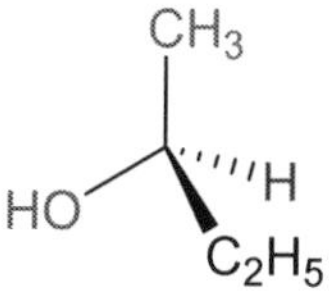

CH₃
HO ⟋⟍ ""H
C₂H₅

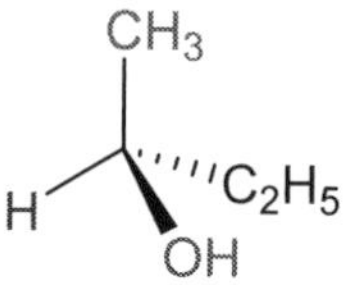

CH₃
H ⟋⟍ ""C₂H₅
OH

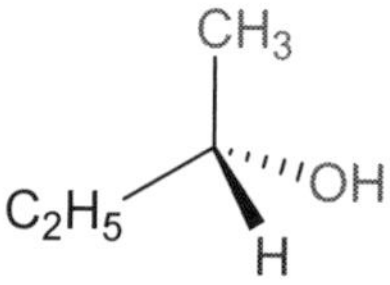

CH₃
C₂H₅ ⟋⟍ ""OH
H

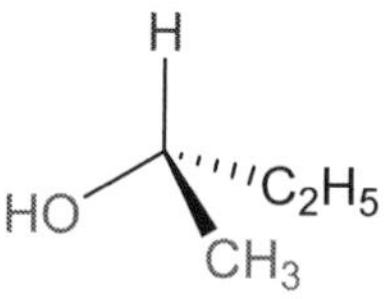

H
HO ⟋⟍ ""C₂H₅
CH₃

(B) Two other ways:

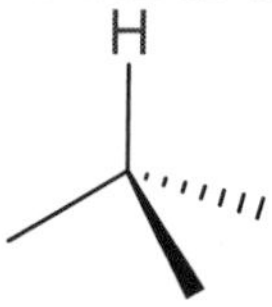

H

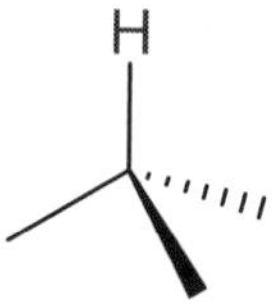

H

Notes

Appendix: Review of Select Topics from Organic Chemistry 1
Fischer projections

In addition to standard dash-wedge notation, a **Fischer projection** can show 3D shape.

In a Fischer projection, horizontal lines: Ⓐ

Vertical lines: Ⓑ

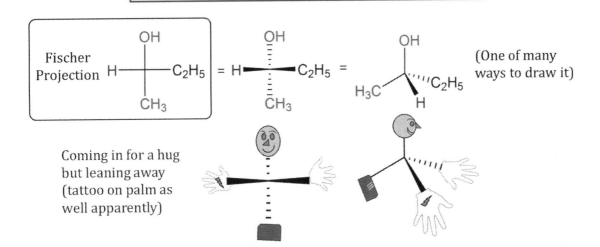

Fischer Projection

$$H-\overset{\displaystyle OH}{\underset{\displaystyle CH_3}{|}}-C_2H_5 = H\blacksquare\overset{\displaystyle OH}{\underset{\displaystyle CH_3}{\vdots}}\blacksquare C_2H_5 =$$

(One of many ways to draw it)

$$H_3C-\overset{OH}{\underset{H}{\overset{|}{C}}}''''C_2H_5$$

Coming in for a hug but leaning away (tattoo on palm as well apparently)

Notes

$$H_3C \overset{\overset{\displaystyle H}{|}}{\underset{\underset{\displaystyle C_2H_5}{|}}{C}} OH$$

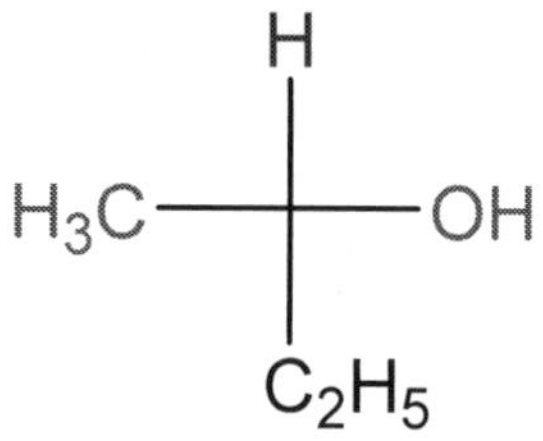

(A) Name:

Convert to wedge and dash structure (several correct representations):

(B)

Fischer Projection and Wedge/Hashed Lines for (*R*)-3-methylheptane:

(C)

Notes

Diastereomers:

A compound with 'n' chiral centers can have up to 2^n stereoisomers.

Consider 3-chloro-2-butanol, which has Stereocenters at C2 and C3:

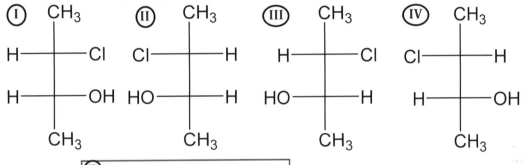

I and II = (B)

III and IV = (C)

Any other pair = (D)

Notes

Meso Compounds:

Ⓐ

So compounds I and II are ACHIRAL!

Consider 2,3-dichlorobutanol, which has Stereocenters at C2 and C3:

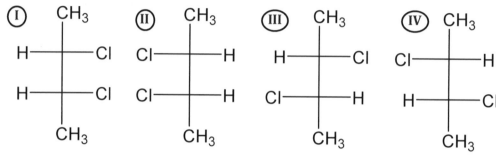

I and II =   Ⓑ

III and IV = Ⓒ   Any other pair =   Ⓓ

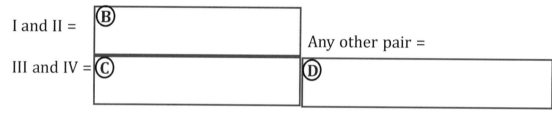

<u>*Notes*</u>

Bimolecular reaction (S_N2, E2) are favored by high concentration of good Nu or strong B

good Nu, weak B   (i.e., I⁻, Br⁻, HS⁻, NH_3): Ⓐ

good Nu, strong B (i.e., HO⁻, EtO⁻, H_2N^-): Ⓑ

poor Nu, strong B (i.e., ^{t}BuO⁻, bulky):   Ⓒ

Substrate can also help determine E2 versus S_N2
   1° RX   no bulky base: Ⓓ

   BULKY base (i.e., ⁻O^{t}Bu): Ⓔ
   2° RX S_N2 **and E2**;
            the stronger/bulkier the base, the more **E2**
   3° RX   **E2 only**

Unimolecular reactions (S_N1, E1) are favored when neither a good nucleophile nor a strong base

---

Notes

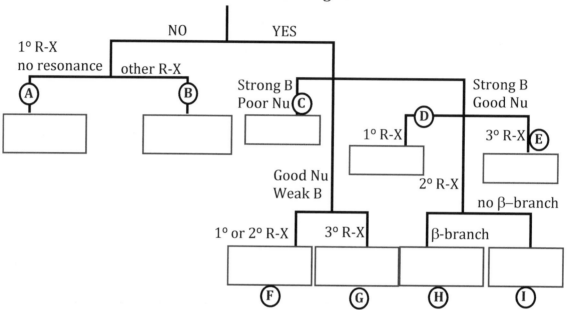

Is there a Good Nu or Strong B?

NO   YES

1° R-X
no resonance   other R-X

A

B

Strong B
Poor Nu C

Strong B
Good Nu

D

1° R-X

3° R-X E

2° R-X

Good Nu
Weak B

no β–branch

1° or 2° R-X   3° R-X   β-branch

F   G   H   I

Notes

Appendix: Review of Select Topics from Organic Chemistry 1
Substitution and elimination practice

Example. Determine whether each reaction will proceed predominantly via S_N1, S_N2, E1, E2, or some combination thereof, and show the product(s)

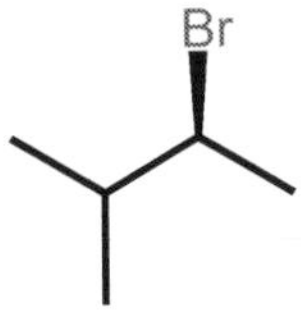

$NaOCH_3$ →

(A)

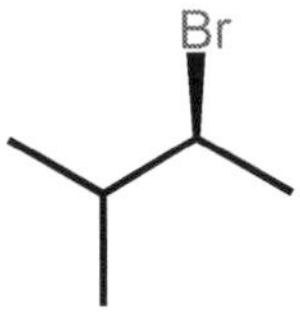

$HOCH_3$ →

(B)

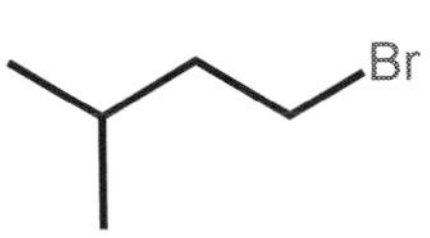

$HOCH_3$ →

(C)

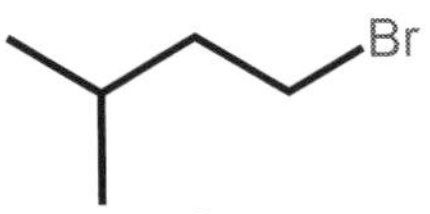

NaOH →

(D)

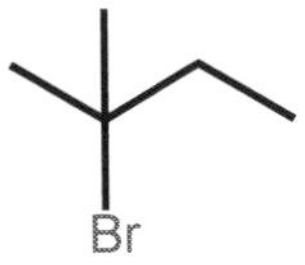

$NaOCH_3$ →

(E)

Notes

Example. Predict whether each proceeds via S_N1, S_N2, E1, or E2, and draw the major product (show stereochemistry where applicable).

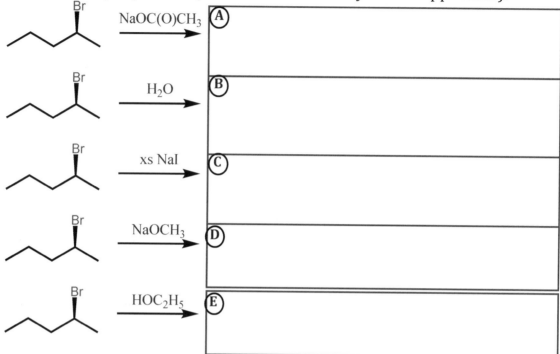

Notes

Example. Predict the major product of each reaction, showing stereochemistry where applicable.

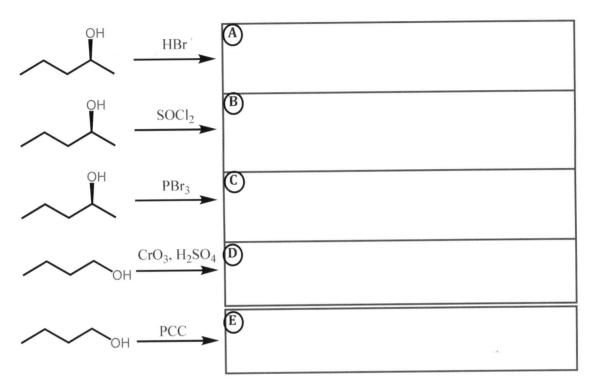

Notes

Example. Predict the major product of each reaction, showing stereochemistry where applicable.

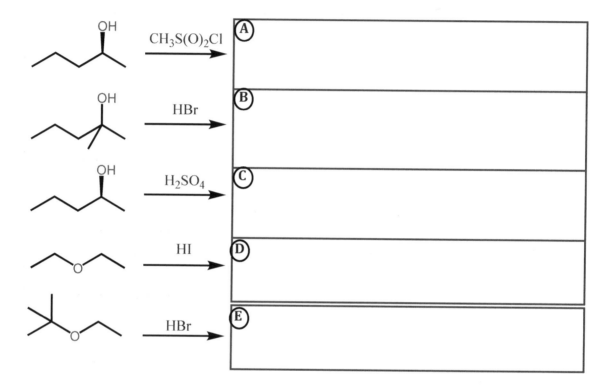

OH, CH₃S(O)₂Cl	Ⓐ
OH, HBr	Ⓑ
OH, H₂SO₄	Ⓒ
HI	Ⓓ
HBr	Ⓔ

Notes

Example. Predict the major product of each reaction, showing stereochemistry where applicable.

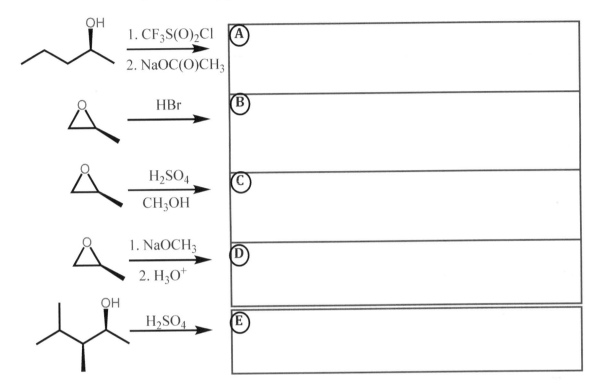

Notes

Appendix: Review of Select Topics from Organic Chemistry 1
Alkene reaction practice

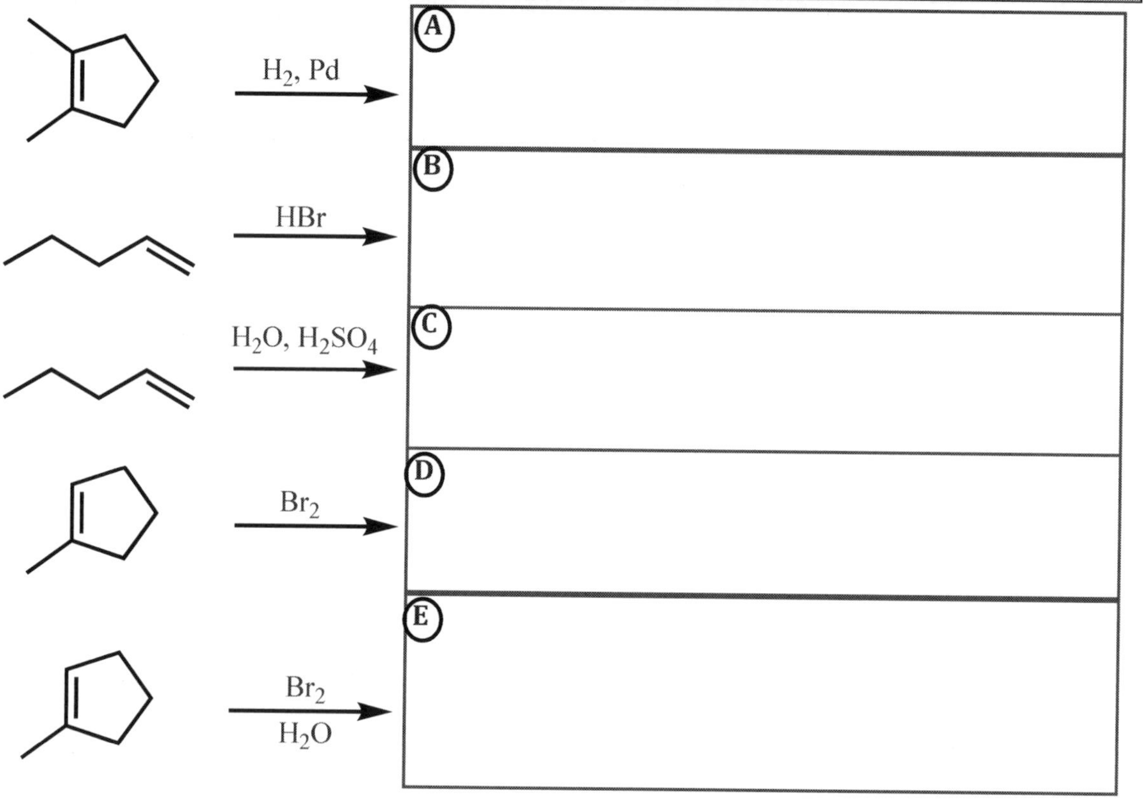

H$_2$, Pd → (A)

HBr → (B)

H$_2$O, H$_2$SO$_4$ → (C)

Br$_2$ → (D)

Br$_2$, H$_2$O → (E)

Notes

Appendix: Review of Select Topics from Organic Chemistry 1
Alkene reaction practice

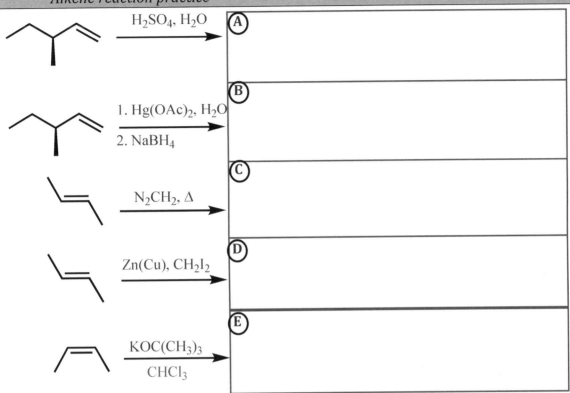

A. H_2SO_4, H_2O

B. 1. $Hg(OAc)_2$, H_2O
   2. $NaBH_4$

C. N_2CH_2, Δ

D. $Zn(Cu)$, CH_2I_2

E. $KOC(CH_3)_3$
   $CHCl_3$

Notes

Alkene reaction practice

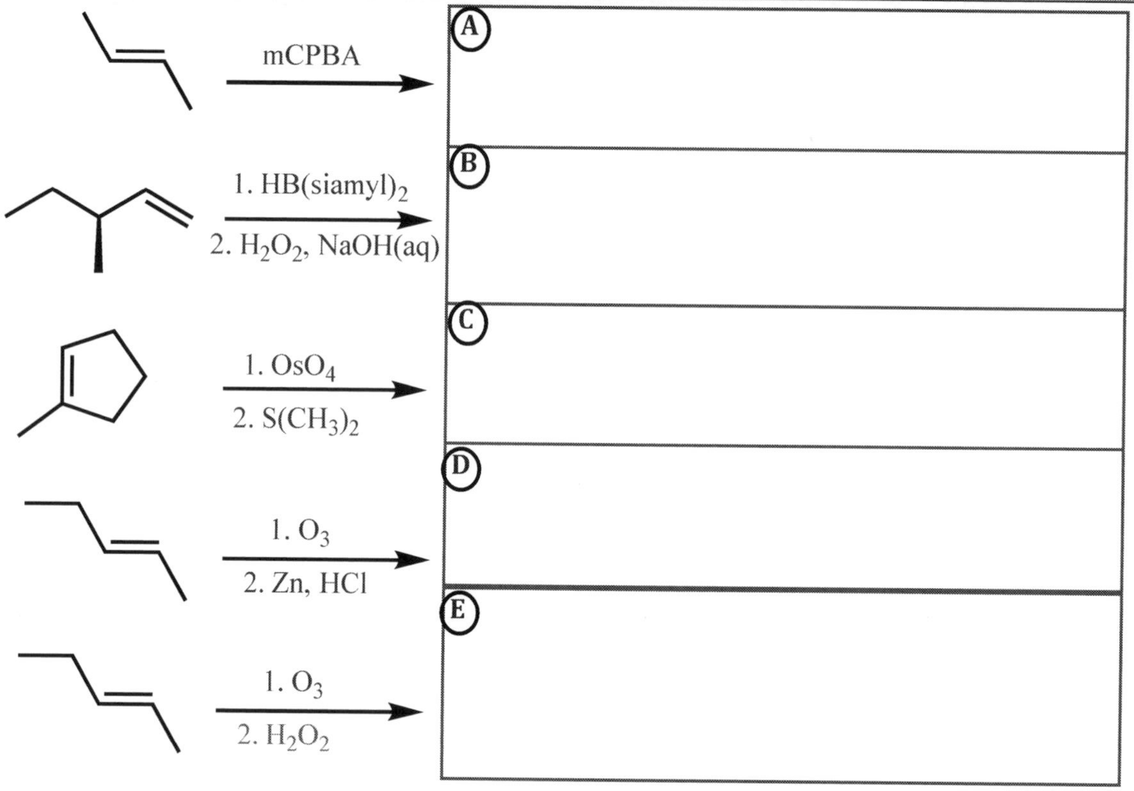
mCPBA
A
1. HB(siamyl)2
2. H2O2, NaOH(aq)
B
1. OsO4
2. S(CH3)2
C
1. O3
2. Zn, HCl
D
1. O3
2. H2O2
E

Notes

INDEX